Nematodes - Ecology, Adaptation and Parasitism

Edited by Soumalya Mukherjee and Sajal Ray

Published in London, United Kingdom

Nematodes - Ecology, Adaptation and Parasitism
http://dx.doi.org/10.5772/intechopen.110977
Edited by Soumalya Mukherjee and Sajal Ray

Contributors
Mohammad Asif, P. Fazul Rahaman, Lalson Wesly Johnson, Rajaswaminathan Vairavan, Venkadesh Ganesan, Gurram Mallikarjun, Katakam Rupini Krishna, Arley Rey Páez, Mohammed Bukar Aji, Himani Sharma, Ashok Kumar Chaubey

First published in London, United Kingdom, 2024 by IntechOpen
IntechOpen is the global imprint of INTECHOPEN LIMITED, registered in England and Wales, registration number: 11086078, 5 Princes Gate Court, London, SW7 2QJ, United Kingdom

British Library Cataloguing-in-Publication Data
A catalogue record for this book is available from the British Library

Additional hard and PDF copies can be obtained from orders@intechopen.com

Nematodes - Ecology, Adaptation and Parasitism
Edited by Soumalya Mukherjee and Sajal Ray
p. cm.
Print ISBN 978-1-83769-941-4
Online ISBN 978-1-83769-940-7
eBook (PDF) ISBN 978-1-83769-942-1

Meet the editors

Soumalya Mukherjee obtained his MSc in Zoology from the West Bengal State University, India. He was selected for the DST INSPIRE Fellowship (both JRF and SRF) from the Department of Science and Technology, Government of India. He received his Ph.D. from the University of Calcutta, India. Dr. Mukherjee published several full-length research articles and review articles in peer-reviewed scientific journals and books of international repute and presented his research findings at various international and national conferences and symposia. He has also served as a book editor. He has been actively engaged in the research of immunotoxicological effects of environmental pollutants on terrestrial and aquatic invertebrates. Previously, he was Assistant Professor of Zoology under the Directorate of Open and Distance Learning at the University of Kalyani, India. Dr. Mukherjee is currently an assistant professor and head of the Department of Zoology, Brahmananda Keshab Chandra College, affiliated with West Bengal State University, India.

Sajal Ray received his MSc and MPhil in Zoology and Environmental Science, respectively, from the University of Calcutta. He obtained a Ph.D. from Jadavpur University, India. His thesis reported the immunotoxicity of pesticides in an economically important snail in India. As an awardee of the Fogarty Visiting Fellowship, Dr. Ray carried out his postdoctoral research in cardiac pathology at the National Institutes of Health, USA. His research interest is studying the immunological responses of molluscs, sponges, crabs, earthworms, and starfishes exposed to pollutants. His team is engaged in understanding the evolutionary mechanism of immunity in phylogeny. He has presented his research at various conferences, including the World Congress of Malacology in Washington DC. Dr. Ray, currently a professor in the Department of Zoology at Calcutta University, India, has been teaching zoology for nearly 30 years at the postgraduate level.

Contents

Preface

The geographical distribution of nematodes is cosmopolitan, encompassing both parasitic and free-living forms. They typically have a worm-like appearance and are extremely adaptive and resilient to a variety of environmental challenges and stresses. They are capable of causing a wide range of diseases in different organisms, including humans. Numerous nematodes are accountable for causing damage in the agricultural sector.

Nematodes present in the soil play an essential role in the functioning of the soil ecosystem. They are excellent indicators of soil health and soil recovery and are found in all habitats, including contaminated and biologically unsafe ones. In addition to maintaining soil biodiversity, they aid in nutrient cycling, aggregation of soil, decomposition, sequestration of carbon, and controlling pests. For the effective management of ecosystems and agriculture in a sustainable manner, it is imperative to comprehend their diversity, interrelationships, and impacts on plants in a holistic way.

In the context of anthropogenic disturbances like mining activities, the related book chapter emphasizes the significance of soil nematodes as critical markers of ecological health. Nematodes in the soil are crucial for the mineralization of nutrients and provide substantial details on the dynamics of the soil food web. A thorough comprehension of the organization of nematode communities can improve their potential as universal indicators for determining the availability of food resources and characterizing diverse habitat types. This work has created opportunities for further research utilizing nematodes as models in coal mine environments. This study paves the way for a more thorough comprehension of how human activity affects soil ecosystems and the possibilities for ecological restoration initiatives.

A special kind of insect parasitic nematodes known as entomopathogenic nematodes is frequently employed in biological pest management. In the related chapter, the authors discuss the impacts of ecology, behavior, symbiosis, and parasitism of entomopathogenic nematodes on their capacity to cause disease in natural conditions. The chapter makes a significant contribution to the knowledge of entomopathogenic nematodes by highlighting their ecological significance and possible effects on sustainable pest management practices.

Plant parasitic nematodes have several remarkable features, particularly their parasitic adaptation that aids in surviving harsh environmental conditions. Moreover, they interact with other pathogenic microorganisms to cause intricate disease in plants. It is crucial to comprehend the parasitic mechanism, feeding behavior, adaptability, host parasitic interaction, and identification of associated microbiota of plant parasitic nematode to manage them effectively within the biological control system.

The use of biopesticides is becoming more and more popular because chemical nematicides against plant parasitic nematodes have some disadvantages. A viable formulation must be created and in-depth research on the virulence activity of entomopathogenic nematodes in in vitro conditions against plant parasitic nematodes is needed to employ them in the field.

To promote the development of environmentally safe compounds, the concept of biological nematicide is extremely appropriate and research in this area should be encouraged. Nematologists should adopt a more methodical approach to biocontrol and should also become more conversant regarding the ecology of soil microbes. In the long run, the designing of nematicides using organic resources may even turn out to be a better and more effective option than synthetic nematicides. However, more field investigations need to be done to evaluate the performance of these materials in actual situations. Furthermore, in organic and environmentally friendly aquaculture, modern technology must be amalgamated with traditional control and management practices to minimize the negative impacts on non-target organisms.

This book, *Nematodes – Ecology, Adaptation and Parasitism*, is a compilation of research articles on the frontiers of nematode biology, including information on the ecological traits, adaptability, and parasitic nature of the Phylum Nemathelminthes. The authors examine the differences in nematode community structures between contaminated and natural environments in terms of diversity, abundance, and bio-indicative potential. The book addresses a variety of topics, including the ecological significance of entomopathogenic nematodes, their potential effects on sustainable pest control management, the parasitic adaptability of plant parasite nematodes, and their interactions with other microorganisms. It also discusses how to manage plant parasitic nematodes by using a new generation of chemical and herbal nematicides. This book is a useful resource for scientists, experts, and students pursuing research on various aspects of nematology.

Soumalya Mukherjee
Department of Zoology,
Brahmananda Keshab Chandra College,
West Bengal, India

Sajal Ray
Aquatic Toxicology Laboratory,
Department of Zoology,
University of Calcutta,
West Bengal, India

Chapter 1

Beneath the Surface: Comparative Diversity, Abundance, and Bio-Indicative Capabilities of Nematode Community Structure in Natural and Disturbed Habitats

Mohammad Asif and P. Fazul Rahaman

Abstract

This present study underscores the significance of soil nematodes as indicators of ecological health, particularly in areas affected by human activities like mining. Soil nematodes, classified into five trophic groups, play vital roles in nutrient cycling and provide insights into soil food web dynamics. Various nematode community indices, including the maturity index (MI) and trophic diversity index (TDI), are crucial for assessing soil food web diversity. Nematodes exhibit adaptability across diverse soil environments, from pristine to highly disturbed habitats, making them responsive indicators of environmental changes. Understanding nematode community structure enhances their potential as global indicators for assessing food resource availability and habitat characterization. This study compares three different zones based on anthropogenic disturbances in the coal mining region of Sonebhadra, Uttar Pradesh, representing undisturbed, moderately disturbed, and intensely disturbed sites. By analyzing nematode communities and trophic group abundance, the study assesses soil ecosystems across these zones. Human activities, especially mining, significantly impact soil nematode diversity and ecosystem health. Transition from natural forests to mining sites leads to shifts in nematode communities and species diversity, with intermediate disturbance fostering increased species diversity. Maturity index values reflect ecosystem maturity, with undisturbed and moderately disturbed zones indicating structured ecosystems, while highly disturbed zones represent degraded conditions. Faunal profiles mirror these findings, indicating shifts in decomposition pathways. This study highlights the potential of nematodes as indicators for environmental monitoring and quality assessment in coal mine areas. Further research on individual nematode species can inform biodiversity modeling and contribute to more effective ecological restoration efforts.

Keywords: soil nematodes, trophic groups, mining activities, faunal profiles, decomposition pathways, biodiversity modeling, ecological restoration

1. Introduction

Nematodes have been categorized into five primary trophic groups [1, 2]. They play a significant role in nutrient mineralization through energy decomposition pathways [3–6] and offer valuable insights into the structure and function of soil food webs [1, 7–9]. Nematode community indices, such as the maturity index (MI) based on c-p scaling, the plant parasitic index (PPI) (weighted mean of c-p values of plant-parasitic nematodes), species richness, evenness, the ratio of bacterivore to fungivore nematodes, and the trophic diversity index (TDI), are routinely employed to assess soil food web diversity conditions [10–20]. Furthermore, various faunal food web indices for nematodes, such as enrichment (EI), structure (SI), and decomposition channel index (CI), based on the relative weighted abundance of nematode c-p guilds, provide information on enrichment, trophic connections, structure, and prevailing decomposition in a food web [1].

Nematodes are valuable indicators as they inhabit soils ranging from pristine to highly disturbed habitats and are represented in almost all trophic groups of the soil food web. They respond quickly and specifically to environmental changes. A comprehensive understanding of nematode community structure for assessing food resource availability and characterizing different habitats would enhance the indicative capability of nematodes on a global scale [1, 21].

Although we have limited knowledge about the influence of vegetation and abiotic parameters on nematode fauna, drawing trends can be challenging due to geological, climatological, and methodological differences among various sites or studies [22]. Analyzing the soil nematode community can reveal differences between habitats [19, 21].

In the present study, three different zones with varying degrees of anthropogenic disturbances in the coal mine region of district Sonebhadra, Uttar Pradesh, were compared. Various parameters of the nematode community were used to assess the prevailing soil ecosystem in these zones. Depending on the nature and frequency of disturbances (physical, chemical, or biological), active mining land and wild forests (potential future mining sites) represent two extremes in terms of ecosystem structure and stability. The three zones identified based on the degree of human interference were: undisturbed hillocks or potential future mining sites (Zone 'A'), moderately disturbed and managed habitat represented by abandoned reclaimed land after mining (Zone 'B'), and intensively disturbed habitat as the active mining site (Zone 'C').

The objectives of the study were to compare these three selected habitats using a combination of nematode community indices and trophic group abundance and to examine substrate characteristics through nematode community measures using multivariate analyses.

2. Description of site and protocol of study

See **Figure 1.**

The coal mine region was divided broadly into three differentiating sites on gradient of anthropogenic disturbances.

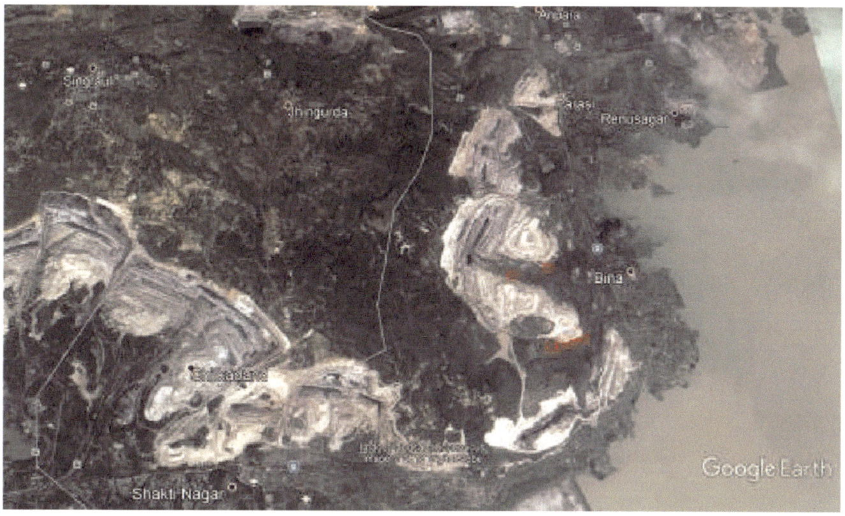

A

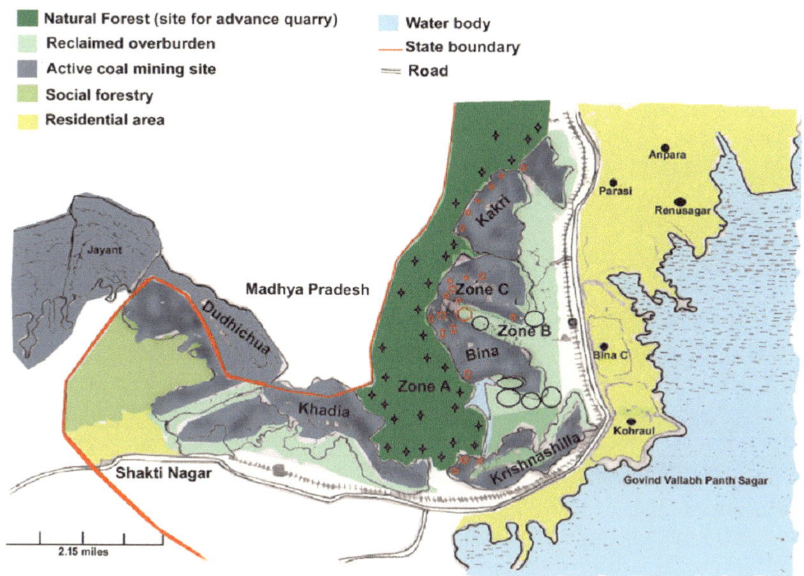

B

Figure 1.
A. Satellite images of the coal mine region. B. Topographic map of the study region showing three different zones. Zone 'A' = green color, Zone 'B' = light green color (samples were collected from median age coalmine spoils from this zone), Zone 'C' = dark blue color.

Zone 'A': This area represented the unmined hills covered with wild vegetation and with least or no anthropogenic activities. Samples were collected on altitudinal gradients from foothill to top. The Zone lies between the geographic coordinates 24° 11′57.73″N, 82° 44′31.95″E to 24° 08′49.06″N, 82° 43′33.85″E (**Figure 2A** and **B**).

Figure 2.
A, B: Snapshots of Zone 'A'.

Zone 'B': The area represented reclaimed overburden of about 21 years of age. For the proper management of spoils, anthropogenic inference occurs time to time. Samples were collected on an altitudinal gradient from foot to top of overburden. This Zone lies between the geographic coordinates 24° 10′32.99″N, 82° 45′51.29″E to 24° 07′40.38″N, 82° 48′32.51″E (**Figure 3A** and **B**).

Zone 'C': The area represented the highly disturbed zone with an extremely nutrient-poor environment. Samples were collected on both altitudinal and horizontal basis. The samples were collected between the geographic coordinates 24° 10′57.92″N, 82° 45′07.83″E to 24° 07′47.38″N, 82° 43′24.89″E (**Figure 4A** and **B**). On horizontal basis, the active coal mining region was considered as source of disturbance, and the samples were collected in a circumference of 400 m diameter.

Figure 3.
A, B: Snapshots of Zone 'B'.

3. Materials and methods

Sampling for Zone 'A': Soil sampling was done on an altitudinal gradient from the foot to the top on undisturbed hills. Samples were taken 15–25 cm below the ground level.

Sampling for Zone 'B': Soil sampling was done from top to bottom and from the bench of the coal mine spoil.

Sampling for Zone 'C': Samples were taken at a 15 m distance along the altitudinal gradient starting from ground level (about 160 m above sea level) to the top (about 400 m high).

A

B

Figure 4.
A, B: Snapshots of Zone 'C'.

Sampling was done using steel corer of 5 cm diam. For each Zone 30 samples were collected with each sample being a composite of five samples. Samples were then stored in plastic bags and transported to the lab for chemical analysis and nematode identification. The climatic and edaphic information were also recorded (**Table 1**). Geographic data like geographic coordinates, altitude, and temperature were recorded *via* the navigation device GARMIN -eTREX- 20.

Nematode species	Functional guild	Mean ± SD (abundance)		
Bacterivores		**Zone 'A' Undisturbed forest (site for advance quarry)**	**Zone 'B' Spoil of median age**	**Zone 'C' Active site of mining**
Achromadora indica	Ba_2	10.56 ± 3.54	28.24 ± 21.24	0.0 ± 0.0
Acrobeles cylindricus	Ba_2	14.39 ± 16.54	20.12 ± 15.54	0.0 ± 0.0
Acrobeloides conoidis	Ba_2	10.34 ± 7.45	12.45 ± 8.64	0.48 ± 0.75
Acrobeloides nanus	Ba_2	18.15 ± 2.56	15.10 ± 12.10	0.0 ± 0.0
Acrobelophis minimus	Ba_2	5.24 ± 3.21	0.63 ± 2.25	0.0 ± 0.0
Alaimus primitivus	Ba_3	18.35 ± 32.30	30.34 ± 44.53	0.0 ± 0.0
Amphidelus sylvaticus	Ba_3	16.3 ± 15.53	13.54 ± 8.64	0.0 ± 0.0
Cephalobus cubaensis	Ba_2	10.4 ± 6.24	7.52 ± 10.25	0.0 ± 0.0
Cephalobus parvus	Ba_2	10.86 ± 7.25	7.81 ± 5.23	0.0 ± 0.0
Ceratoplectus armatus	Ba_2	9.26 ± 4.23	19.83 ± 25.3	0.0 ± 0.0
Chiloplacus symmetricus	Ba_2	1.45 ± 7.59	0.25 ± 1.25	0.0 ± 0.0
Chiloplectus indicus	Ba_2	0.50 ± 2.76	0.65 ± 2.54	0.0 ± 0.0
Eucephalobus oxyuroides	Ba_2	16.68 ± 18.85	18.25 ± 12.24	0.0 ± 0.0
Geomonhystera glandulata	Ba_2	12.24 ± 8.35	8.53 ± 12.53	0.0 ± 0.0
Halicephalobus gingivalis	Ba_1	0.00 ± 0.00	0.94 ± 3.65	0.0 ± 0.0
Mesorhabditis vernalis	Ba_1	13.51 ± 8.56	22.45 ± 9.45	0.36 ± 0.67
Mesorhabditis minuta	Ba_1	8.45 ± 6.45	15.53 ± 12.67	0.48 ± 0.84
Monhystera gracilis	Ba_2	11.84 ± 6.85	15.28 ± 14.53	0.0 ± 0.0
Monhystrella kerryi	Ba_2	12.54 ± 6.44	9.28 ± 5.68	0.0 ± 0.0
Oscheius vulvastriatus sp .n.	Ba_1	0.00 ± 0.00	8.22 ± 3.25	0.0 ± 0.0
Panagrellus dubius	Ba_1	3.15 ± 1.23	1.75 ± 0.57	0.0 ± 0.0
Panagrolaimus dendroctoni	Ba_2	15.83 ± 6.86	8.75 ± 6.55	0.0 ± 0.0
Plectus geophilus	Ba_2	16.84 ± 18.24	18.49 ± 20.24	0.0 ± 0.0
Plectus parvus	Ba_2	7.42 ± 4.24	5.77 ± 14.99	0.0 ± 0.0
Poikilolaimus oxycercus	Ba_1	2.15 ± 0.24	0.55 ± 2.50	0.0 ± 0.0
Prismatolaimus intermedius	Ba_3	12.46 ± 14.48	16.86 ± 12.86	0.0 ± 0.0
Teratolobus regulus	Ba_3	2.43 ± 1.75	1.25 ± 3.45	0.0 ± 0.0

Nematode species	Functional guild	Mean ± SD (abundance)		
Bacterivores		Zone 'A' Undisturbed forest (site for advance quarry)	Zone 'B' Spoil of median age	Zone 'C' Active site of mining
Tylocephalus primitivus	Ba_2	3.26 ± 2.36	3.86 ± 1.67	0.0 ± 0.0
Wilsonema bangaloreiensis	Ba_2	3.35 ± 1.75	3.55 ± 5.43	0.0 ± 0.0
Zeldia punctata	Ba_2	2.72 ± 2.24	0.74 ± 2.24	0.0 ± 0.0

Table 1.
Abundance of bacterivore species in three zones of coal mine region of district Sonebhadra Uttar Pradesh.

4. Results and observations

4.1 Nematode abundance

Zone 'A': Nematode abundance was found high in Zone 'A' and mean value of 1748.63 ± 163.18 (1576–1859) nematodes were found in 100 g of samples.

Zone 'B': Nematode abundance was low in Zone 'B' comparably to Zone 'A' and mean value of 983.8 ± 120.2 (748.57–1168.74) nematodes were found in 100 g of samples.

Zone 'C': Nematode abundance was negligible in Zone 'C', which was the active mining Zone and mean of 0.42 ± 0.37 (0.08–0.42) nematodes isolated from 100 g of samples.

4.2 Nematode species diversity

Zone 'A': A total of 119 species belonging to 107 genera, 40 families and eight orders were recorded from Zone 'A'. The dominant species were *Aphelenchoides composticola, Aphelenchus avenae, Dorylaimoides constrictus, Labronema baqrii, Labronema deoriaensis, Laimydorus baldus, Thornenema mauritianum, Tripyla glomerans, Tripyla arenicola, Mylonchulus contractus, Mylonchulus vasis, Sporonchulus ibitiensis, Aporcelaimellus tropicus.*

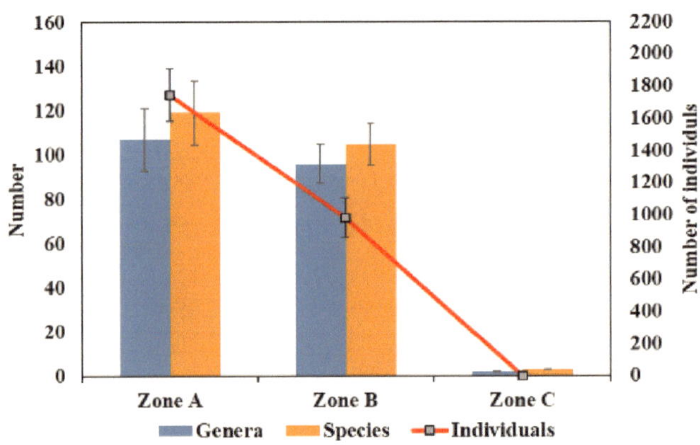

Figure 5.
Number of species, genera and individuals observed in three different zones of coal mine region.

Zone 'B': Zone 'B' recorded 105 species belonging to 96 genera, 39 families and eight orders. The number and dominance were quite comparable with Zone 'A'.

Zone 'C': Only three species belonging to two genera, one family and one order were observed at Zone 'C' (**Figure 5**).

4.3 Nematode trophic structure

In Zone 'A' there was much diversity of omnivores and predators in comparison to bacterivores and in Zone 'B' bacterivores and fungivores nematodes were the abundant group over omnivores and predators and Zone 'C' had only three bacterivore species.

4.3.1 Zone 'A'

4.3.1.1 Trophic groups

Bacterivores (bacteriophagous): Of the 28 bacterivore species recorded (**Table 2**), five most abundant species were *Acrobeloides nanus, Plectus geophilus, Eucephalobus oxyuroides, Amphidelus sylvaticus and Acrobeles cylindricus*.

Nematode species	**Functional guild**	**Mean ± SD (abundance)**		
Fungivores		**Zone 'A' Undisturbed forest (Site for advance quarry)**	**Zone 'B' Spoil of Median Age**	**Zone 'C' Active site of mining**
Aglenchus agricola	Fu_2	16.43 ± 6.05	16.48 ± 8.45	0.0 ± 0.0
Aphelenchus avenae	Fu_2	32.04 ± 18.64	12.6 ± 10.35	0.0 ± 0.0
Aphelenhoides besseyi	Fu_2	36.13 ± 20.13	18.74 ± 16.36	0.0 ± 0.0
Axonchium thoubalicum	Fu_5	9.40 ± 3.56	2.25 ± 1.75	0.0 ± 0.0
Basiriotyleptus minimus	Fu_2	8.95 ± 4.31	6.25 ± 4.21	0.0 ± 0.0
Belondira apitica	Fu_5	10.42 ± 4.10	5.40 ± 2.50	0.0 ± 0.0
Daladenus wilsoni	Fu_2	3.67 ± 1.58	13.43 ± 6.48	0.0 ± 0.0
Diptherophora alami	Fu_3	12.10 ± 4.20	4.61 ± 2.80	0.0 ± 0.0
Ditylenchus dipsaci	Fu_2	3.26 ± 2.05	11.4 ± 6.23	0.0 ± 0.0
Dorylaimellus indicus	Fu_5	16.82 ± 6.42	8.64 ± 3.56	0.0 ± 0.0
Dorylaimoides constrictus	Fu_2	39.27 ± 20.46	13.45 ± 6.45	0.0 ± 0.0
Filenchus mircodorus	Fu_2	22.30 ± 10.46	16.43 ± 8.23	0.0 ± 0.0
Filenchus vulgaris	Fu_2	18.20 ± 9.65	15.43 ± 6.28	0.0 ± 0.0
Leptonchus granulosus	Fu_2	4.85 ± 2.24	5.44 ± 3.24	0.0 ± 0.0
Roqueus indicus	Fu_5	4.64 ± 3.41	5.53 ± 2.54	0.0 ± 0.0
Tylencholaimellus acutus	Fu_4	28.41 ± 14.42	12.34 ± 6.80	0.0 ± 0.0
Tyleptus projectus	Fu_2	7.23 ± 3.54	10.23 ± 4.25	0.0 ± 0.0

Table 2.
Abundance of fungivores species in three zones of coal mine region of district Sonebhadra Uttar Pradesh.

Fungivores (mycophagous): Seventeen fungivore species were recorded (**Table 3**), among which the five most abundant ones were *Dorylaimoides constrictus, Aphelenhoides composticola, Aphelenchus avenae, Tylencholaimellus acutus* and *Filenchus mircodorus.*

Herbivores (plant parasites): Of the 22 herbivore species (**Table 4**), five most abundant ones were *Longidorus elongates, Basiria abberans, Longidoroides longiurus, Tylenchus arcuatus* and *Tylenchorynchus striatus.*

Omnivores: Five most abundant species among the 25 omnivore species (**Table 5**) were *Allodorylaimus irritans, Thornenema mauritianum, Labronema deoriaensis, Mesodorylaimus indicus* and *Mesodorylaimus subtiloides.*

Predators: Of a total of 26 predator species (**Table 6**), the five most abundant ones were recorded as *Tripyla arenicola Mylonchulus viasis Tripyla glomerans Mylonchulus contractus* and *Laimydorus baldus.*

Nematode species	Functional guild	Mean ± SD (abundance)		
Herbivores		**Zone 'A' Undisturbed forest (site for advance quarry)**	**Zone 'B' Spoil of median age**	**Zone 'C' Active site of mining**
Basiria abberans	H_2	16.80 ± 6.46	7.25 ± 4.23	0.0 ± 0.0
Boleodorus filiformis	H_2	9.35 ± 4.05	3.36 ± 2.35	0.0 ± 0.0
Coslenchus areolatus	H_2	6.87 ± 5.64	1.53 ± 2.50	0.0 ± 0.0
Helicotylenchus dihystera	H_3	9.64 ± 4.38	12.68 ± 5.85	0.0 ± 0.0
Hemicriconemoides communis	H_3	15.38 ± 8.36	6.86 ± 4.25	0.0 ± 0.0
Hemicycliophora corbetti	H_3	1.50 ± 2.35	1.02 ± 2.21	0.0 ± 0.0
Hoplolaimus indicus	H_3	2.30 ± 3.66	10.25 ± 6.23	0.0 ± 0.0
Longidoroides longiurus	H_5	13.45 ± 8.36	2.52 ± 0.0	0.0 ± 0.0
Longidorus elongatus	H_5	16.84 ± 10.35	1.08 ± 2.50	0.0 ± 0.0
Malenchus nanellus	H_2	4.68 ± 3.45	1.24 ± 3.65	0.0 ± 0.0
Merlinius brevidens	H_3	3.87 ± 2.36	1.25 ± 0.58	0.0 ± 0.0
Neopsilenchus affinis	H_2	2.48 ± 2.01	1.20 ± 2.36	0.0 ± 0.0
Paratylenchus similis	H_2	9.85 ± 6.46	12.26 ± 6.58	0.0 ± 0.0
Pratylenchus brachyurus	H_2	10.64 ± 6.68	0.0 ± 0.0	0.0 ± 0.0
Psilenchus mixus	H_3	4.65 ± 3.34	0.56 ± 1.14	0.0 ± 0.0
Rotylenchus mircrostylus	H_3	6.80 ± 4.88	0.26 ± 2.24	0.0 ± 0.0
Scutellonema brachyurum	H_3	1.54 ± 2.34	0.50 ± 1.54	0.0 ± 0.0
Scutylenchus hexincisus	H_3	2.38 ± 1.36	0.35 ± 2.25	0.0 ± 0.0
Trichodorus borai	H_4	8.62 ± 4.38	1.24 ± 2.52	0.0 ± 0.0
Tylenchorynchus striatus	H_3	11.56 ± 8.44	10.23 ± 6.45	0.0 ± 0.0
Tylenchus arcuatus	H_3	12.56 ± 6.35	5.86 ± 3.54	0.0 ± 0.0
Xiphinema americanum	H_5	6.24 ± 9.25	9.25 ± 5.24	0.0 ± 0.0

Table 3.
Abundance of herbivore species in three zones of coal mine region of district Sonebhadra Uttar Pradesh.

Nematode species	Functional guild	Mean ± SD (abundance)		
Omnivores		**Zone 'A' Undisturbed forest (site for advance quarry)**	**Zone 'B' Spoil of median age**	**Zone 'C' Active site of mining**
Amphidorylaimus flagellicauda	Om_4	6.24 ± 4.36	—	—
Cephalodorylaimus papillatus	Om_4	6.20 ± 5.48	—	—
Crassolaibium gharwaliense	Om_4	16.84 ± 9.65	4.24 ± 2.35	—
Discomyctus cephalatus	Om_4	12.86 ± 9.80	6.33 ± 3.26	—
Dorylaimus stagnalis	Om_4	10.54 ± 5.46	1.56 ± 3.28	—
Enchodelus macrodorus	Om_4	8.64 ± 5.24	0.45 ± 1.56	—
Eudorylaimus silvaticus	Om_4	13.64 ± 8.30	5.26 ± 3.25	—
Eumenicus monhystera	Om_4	8.64 ± 5.45	—	—
Ischiyodorylaimus navus	Om_4	3.58 ± 4.24	—	—
Labronema deoriaensis	Om_4	20.42 ± 12.92	15.50 ± 11.54	—
Labronema baqrii	Om_4	24.56 ± 18.45	18.35 ± 8.62	—
Allodorylaimus irritans	Om_4	31.54 ± 22.56	2.45 ± 3.56	—
Makatinus heynsi	Om_4	7.45 ± 6.52	—	—
Mesodorylaimus subtiloides	Om_4	20.38 ± 11.36	—	—
Mesodorylaimus indicus	Om_4	20.38 ± 16.36	23.56 ± 13.46	—
Microdorylaimus parvus	Om_4	18.45 ± 8.46	8.54 ± 5.64	—
Moshajia idiofora	Om_4	12.45 ± 6.75	8.65 ± 5.60	—
Opisthodorylaimus cavalcantii	Om_4	16.45 ± 9.45	—	—
Oriverutus sundarus	Om_4	19.44 ± 12.46	—	—
Oxydirus ganagticus	Om_5	10.65 ± 6.54	—	—
Paraoxydirus gigas	Om_5	8.64 ± 4.84	—	—
Porodorylaimus sturhani	Om_4	15.46 ± 12.24	—	—
Pungentella porosa	Om_4	9.86 ± 5.68	—	—
Pungentus angulatus	Om_4	12.6 ± 9.83	2.60 ± 1.35	—
Thonus cylindricus	Om_4	14.24 ± 8.38	13.50 ± 8.45	—
Thornenema mauritianum	Om_4	22.54 ± 16.38	15.68 ± 8.54	—

Table 4.
Abundance of omnivores species in three zones of coal mine region of district Sonebhadra Uttar Pradesh.

Nematode species	Functional guild	Mean ± SD (abundance)		
Predators		**Zone 'A' Undisturbed forest (site for advance quarry)**	**Zone 'B' Spoil of median age**	**Zone 'C' Active site of mining**
Actinolaimus armatus	Ca_5	12.22 ± 5.64	4.68 ± 2.52	—
Actus minutus	Ca_4	11.22 ± 6.42	6.86 ± 6.36	—

Nematode species	Functional guild	Mean ± SD (abundance)		
Predators		Zone 'A' Undisturbed forest (site for advance quarry)	Zone 'B' Spoil of median age	Zone 'C' Active site of mining
Aporcelaimellus tropicus	Ca_4	22.60 ± 14.38	15.50 ± 8.46	—
Aporcelaimellus heynsi	Ca_4	12.68 ± 8.42	—	—
Aquatides minutus	Ca_5	4.23 ± 6.69	9.40 ± 6.53	—
Clarkus sheri	Ca_4	9.68 ± 4.38	—	—
Discolaimodes bulbiferus	Ca_5	16.46 ± 10.26	14.84 ± 5.21	—
Discolaimus major	Ca_5	12.45 ± 8.62	15.60 ± 8.64	—
Iotonchus monhystera	Ca_4	10.12 ± 5.36	6.18 ± 2.48	—
Iotonchus indicus	Ca_4	15.99 ± 8.59	4.21 ± 3.26	—
Ironus dentifurcatus	Ca_4	18.62 ± 9.96	8.64 ± 4.42	—
Laimydorus papillatus	Ca_4	16.45 ± 4.66	—	—
Laimydorus baldus	Ca_4	22.68 ± 6.68	15.21 ± 8.62	—
Michonchus rectangularis	Ca_4	14.44 ± 8.64	11.64 ± 6.30	—
Mononchoides longicauda	Ca_1	—	0.25 ± 3.36	—
Mulveyellus jairi	Ca_4	9.66 ± 3.36	1.89 ± 2.06	—
Mylonchulus contractus	Ca_4	23.45 ± 14.64	15.24 ± 11.42	—
Mylonchulus viasis	Ca_4	26.20 ± 18.46	16.54 ± 8.44	—
Neoactinolaimus thornei	Ca_5	6.48 ± 5.36	3.05 ± 2.58	—
Nygellus subclavatus	Ca_4	16.28 ± 6.66	6.26 ± 5.88	—
Nygolaimus timmi	Ca_5	8.35 ± 6.45	2.83 ± 1.26	—
Prionchulus muscorum	Ca_4	8.32 ± 4.32	—	—
Solidedens bisexualis	Ca_5	6.64 ± 7.38	4.21 ± 2.36	—
Sporonchulus ibitiensis	Ca_4	25.68 ± 44.61	10.58 ± 13.60	—
Sporonchulus vagabundus	Ca_4	18.68 ± 12.68	5.36 ± 3.36	—
Tripyla glomerans	Ca_3	24.35 ± 18.38	32.26 ± 22.42	—
Tripyla arenicola	Ca_3	31.24 ± 24.24	20.25 ± 12.24	—

Table 5.
Abundance of predator species in three zones of coal mine region of district Sonebhadra Uttar Pradesh.

	Type Area 'A' Non minning forest	Type Area 'B' Spoil of median age	Type Area 'C' Active site of mining
Genera (total count)	109	104	2
Species (total count)	114	110	3
Individuals	1748.63 ± 163.18 (1576–1859)	983.80 ± 120.24 (748.57–1168.74)	0.42 ± 0.37 (0.08–0.42)

DOI: http://dx.doi.org/10.5772/intechopen.1003686

	Type Area 'A' Non minning forest	Type Area 'B' Spoil of median age	Type Area 'C' Active site of mining
Margalef diversity	16.17 ± 0.35 (15.58–16.79)	15.11 ± 0.59 (14.23–16.23)	—
Menhinick diversity	3.92 ± 0.35 (3.63–3.95)	3.43 ± 0.43 (2.85–3.98)	—
Simpson dominance	0.01 ± 0.00 (0.01–0.01)	0.01 ± 0.00 (0.01–0.01)	—
Inverse Simpson dominance	91.14 ± 3.25 (82.25–96.35)	67.50 ± 7.33 (54.32–76.15)	—
Hill's effective index	98.23 ± 2.43 (97.24–99.89)	75.64 ± 4.90 (65.18–77.89)	—
Shannon diversity	4.58 ± 0.45 (3.97–5.65)	4.32 ± 0.44 (3.54–4.88)	—
Simpson diversity	0.99 ± 0.00 (0.99–0.99)	0.99 ± 0.00 (0.99–0.99)	—
Brillouin diversity	4.51 ± 0.65 (3.85–5.78)	4.19 ± 0.34 (3.50–4.60)	—
Sheldon Index	0.82 ± 0.00 (0.82–0.82)	0.72 ± 0.00 (0.72–0.72)	—
Heip Index	0.82 ± 0.00 (0.82–0.82)	0.72 ± 0.00 (0.72–0.72)	—
Maturity Index (MI)	3.46 ± 0.26 (3.36–3.87)	2.96 ± 0.50 (2.87–3.12)	—
Plant Parasitic Index (PPI)	0.78 ± 0.09 (0.50–1.25)	3.26 ± 0.23 (2.00–3.65)	—
Trophic Diversity Index (TDI)	1.17 ± 0.02 (1.06–1.32)	1.09 ± 0.04 (1.04–1.25)	1.00 ± 0.00 (1.04–1.25)
Nematode channel ratio (NCR %)	0.40 ± 0.02 (0.35–0.72)	0.68 ± 0.02 (0.62–0.73)	—
Structure Index (SI)	68.23 ± 10.20 (58.30–77.43)	45.1 5 ± 5.85 (40.68–52.68)	—
Enrichment Index (EI)	12.62 ± 4.66 (7.65–32.25)	22.62 ± 12.58 (16.53–64.65)	9.54 ± 1.54 (9.50–10.56)
Basal Index (BI)	41.28 ± 12.30 (32.04–62.05)	57.29 ± 8.10 (44.24–67.35)	74.46 ± 3.25 (70.13–78.16)
Channel Index (CI)	0.31 ± 0.40 (0.25–0.84)	0.54 ± 0.42 (0.50–1.24)	—

Table 6.
Ecological parameters of nematode population in three zones of coal mine region of district Sonebhadra Uttar Pradesh.

4.3.2 Zone 'B'

4.3.2.1 Trophic groups

Bacterivores (bacteriophagous): Of the 29 bacterivore species recorded (**Table 2**), five most abundant species were *Alaimus primitivus, Achromadora indica, Mesorhabditis vernalis, Acrobeles cylindricus* and *Ceratoplectus armatus.*

Fungivores (mycophagous): Seventeen fungivore species were recorded (**Table 3**), among which five most abundant ones were *Aphelenhoides composticola, Aglenchus agricola, Filenchus mircodorus, Dorylaimoides constrictus* and *Aphelenchus avenae.*

Herbivores (plant parasites): Of the 21 herbivore species (**Table 4**), five most abundant ones were *Helicotylenchus dihystera, Paratylenchus similis, Hoplolaimus indicus, Tylenchorynchus striatus* and *Xiphinema americanum*.

Omnivores: The five most abundant species among the 15 omnivore species (**Table 5**) were *Mesodorylaimus indicus, Labronema baqrii, Thornenema mauritianum, Labronema deoriaensis* and *Thonus cylindricus*.

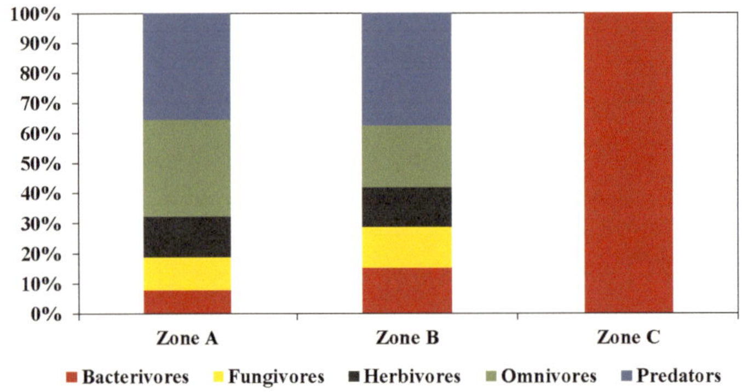

Figure 6.
Percentage biomass of the different trophic groups in the three different zones of coalfield region.

Predators: Of a total of 22 predator species (**Table 6**), the five most abundant ones were recorded as *Tripyla glomerans, Tripyla arenicola, Mylonchulus vasis, Discolaimus major* and *Aprocelaimellus tropicus.*

4.3.3 Zone 'C'

4.3.3.1 Trophic group

Bacterivores: In Zone 'C' only three species were reported from bacteriovores (**Table 2**) *viz., Mesorhabditis vernalis, Mesorhabditis minuta* and *Acrobeloides nanus.*

4.4 Biomass

The relative biomass of different trophic groups (**Figure 6**) varied in three zones and are as follows:

Zone 'A': The highest biomass was of predators, that is, about 36% of the total nematode biomass and the lowest,that is, 8% was constituted by bacterivores. Omnivore species showed 32% biomass followed by herbivores (13%) while fungivores constituted 11% of the total nematode biomass. The total nematode biomass in this Zone was calculated to be 6162.99 μg in 100 g. of soil.

Zone 'B': The nematode biomass demonstrated similar trends in Zone 'B' compared to Zone 'A'. The highest biomass was of predators that is, 37% of the total nematode biomass while herbivores constituted the lowest 13%, omnivores constituted 21% followed by bacterivores (15%) and fungivores (14%) of the total nematode biomass. The total nematode biomass in this Zone was estimated to be 3438.93 μg in 100 g of soil.

Zone 'C': In Zone 'C' there were only three bacterivore species that constitute nematode biomass. The total nematode biomass in this Zone was estimated to be 2.37 μg in 100 g of soil.

5. Diversity indices

Diversity indices and nematode maturity indices were calculated for each zone to assess the diversity of nematode species and maturity of soil ecosystem.

5.1 Margalef index (species richness)

In Zone 'A' the Margalef species richness index was found to be 16.17 ± 0.35 (15.58–16.79). Zone 'B' was calculated to be 15.11 ± 0.59 (14.23–16.23) and in Zone 'C', where only two species were recorded, the index was incalculable (**Table** 7 and **Figure** 7).

5.2 Menhinick index (species richness)

Menhinick diversity, a measure to compare the species richness, was calculated to be 3.92 ± 0.35 (3.63–3.95) in Zone 'A', in Zone 'B' it was found to be 3.43 ± 0.43 (2.85–3.98) and the Menhinick index was incalculable for Zone 'C' (**Table** 7 and **Figure** 7).

Characteristics	Mining sites		
	Zone 'A' Undisturbed forest (site for advance quarry)	Zone 'B' Spoil of median age	Zone 'C' Active site of mining
<2.0 mm (gravel)	9.00 ± 0.92	13.00 ± 1.80	20.00 ± 2.15
2.–0.2 mm (sand)	63.00 ± 3.50	72.00 ± 5.24	88.55 ± 6.54
0.2–0.1 mm (silt)	13.00 ± 1.23	10.00 ± 0.24	3.00 ± 1.26
<0.1 mm (clay)	8.10 ± 1.45	7.80 ± 1.25	3.20 ± 1.56
pH	6.54 ± 0.02	6.82 ± 0.02	5.45 ± 0.26
Bulk density	1.25 ± 0.18	1.26 ± 0.08	1.75 ± 0.16
Natural moisture content (%)	12.20 ± 2.26	11.60 ± 0.28	6.2 ± 0.25
Porosity	46.27 ± 2.25	42.62 ± 2.65	33.65 ± 2.66
Water holding capacity	53.00 ± 0.92	52.00 ± 2.65	24.84 ± 3.25
Soil organic carbon (mg C/g)	3.45 ± 0.024	2.38 ± 0.028	ND
Total soil N (μg N/g)	2456.63 ± 148.34	194.94 ± 25.46	ND
Exchangeable K μg P/g spoil	271.95 ± 6.74	16.96 ± 2.35	ND
Electrical conductivity ms/cm	0.05 ± .01	0.33 ± 0.01	0.65 ± 0.01

Table 7.
Abiotic parameters in three zones of coal mine region of district Sonebhadra Uttar Pradesh.

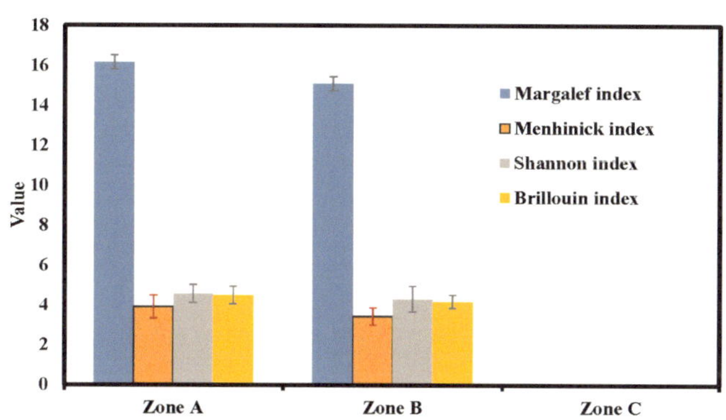

Figure 7.
Comparison of species diversity indices in three selected zones.

Variables	Bac. Sp	Fung. Sp	Herb. Sp	Omi. Sp	Pred. sp.	MI	PPI	TDI	SI	EI	BI	CI	NCR	Tot. Indv.	pH	BD	Porosity	SM	WHC	ToC	TSN	Avail. K	EC
Bac. Sp	1	0.996	0.997	0.872	0.991	0.949	0.999	0.898	0.983	0.705	−0.861	0.884	0.982	0.886	0.988	−0.986	−0.850	0.944	0.996	0.943	0.932	0.534	−0.804
Fung. Sp	0.996	1	1.000	0.910	0.999	0.972	0.999	0.932	0.995	0.643	−0.901	0.920	0.963	0.921	0.998	−0.996	−0.891	0.968	1.000	0.968	0.899	0.603	−0.851
Herb. Sp	0.997	1.000	1	0.905	0.998	0.969	1.000	0.927	0.994	0.653	−0.896	0.915	0.967	0.917	0.997	−0.995	−0.885	0.965	1.000	0.964	0.904	0.593	−0.844
Omi. Sp	0.872	0.910	0.905	1	0.931	0.982	0.892	0.999	0.948	0.269	−1.000	1.000	0.766	1.000	0.937	−0.942	−0.999	0.985	0.914	0.985	0.636	0.879	−0.992
Pred. sp.	0.991	0.999	0.998	0.931	1	0.983	0.996	0.950	0.999	0.602	−0.923	0.939	0.948	0.941	1.000	−0.999	−0.914	0.980	0.999	0.980	0.874	0.644	−0.877
MI	0.949	0.972	0.969	0.982	0.983	1	0.961	0.991	0.991	0.445	−0.978	0.986	0.873	0.987	0.986	−0.988	−0.973	1.000	0.974	1.000	0.770	0.774	−0.951
PPI	0.999	0.999	1.000	0.892	0.996	0.961	1	0.916	0.990	0.675	−0.882	0.903	0.974	0.905	0.994	−0.992	−0.871	0.957	0.999	0.956	0.916	0.569	−0.828
TDI	0.898	0.932	0.927	0.999	0.950	0.991	0.916	1	0.964	0.321	−0.997	1.000	0.800	1.000	0.955	−0.959	−0.995	0.993	0.935	0.993	0.678	0.852	−0.984
SI	0.983	0.995	0.994	0.948	0.999	0.991	0.990	0.964	1	0.562	−0.940	0.955	0.931	0.956	1.000	−1.000	−0.933	0.989	0.996	0.988	0.849	0.681	−0.900
EI	0.705	0.643	0.653	0.269	0.602	0.445	0.675	0.321	0.562	1	−0.247	0.291	0.825	0.295	0.588	−0.576	−0.226	0.432	0.636	0.429	0.914	−0.223	−0.145
BI	−0.861	−0.901	−0.896	−1.000	−0.923	−0.978	−0.882	−0.997	−0.940	−0.247	1	−0.999	−0.751	−0.999	−0.929	0.935	1.000	−0.981	−0.905	−0.981	−0.619	−0.889	0.995
CI	0.884	0.920	0.915	1.000	0.939	0.986	0.903	1.000	0.955	0.291	−0.999	1	0.781	1.000	0.945	−0.950	−0.998	0.989	0.923	0.989	0.654	0.868	−0.989
NCR	0.982	0.963	0.967	0.766	0.948	0.873	0.974	0.800	0.931	0.825	−0.751	0.781	1	0.783	0.942	−0.937	−0.737	0.866	0.961	0.864	0.983	0.367	−0.678
Tot. Indv.	0.886	0.921	0.917	1.000	0.941	0.987	0.905	1.000	0.956	0.295	−0.999	1.000	0.783	1	0.946	−0.951	−0.997	0.989	0.925	0.990	0.658	0.866	−0.988
pH	0.988	0.998	0.997	0.937	1.000	0.986	0.994	0.955	1.000	0.588	−0.929	0.945	0.942	0.946	1	−1.000	−0.921	0.983	0.998	0.983	0.866	0.657	−0.886
BD	−0.986	−0.996	−0.995	−0.942	−0.999	−0.988	−0.992	−0.959	−1.000	−0.576	0.935	−0.950	−0.937	−0.951	−1.000	1	0.926	−0.986	−0.997	−0.986	−0.858	−0.669	0.892
Porosity	−0.850	−0.891	−0.885	−0.999	−0.914	−0.973	−0.871	−0.995	−0.933	−0.226	1.000	−0.998	−0.737	−0.997	−0.921	0.926	1	−0.976	−0.895	−0.977	−0.602	−0.899	0.997
SM	0.944	0.968	0.965	0.985	0.980	1.000	0.957	0.993	0.989	0.432	−0.981	0.989	0.866	0.989	0.983	−0.986	−0.976	1	0.971	1.000	0.760	0.783	−0.955
WHC	0.996	1.000	1.000	0.914	0.999	0.974	0.999	0.935	0.996	0.636	−0.905	0.923	0.961	0.925	0.998	−0.997	−0.895	0.971	1	0.970	0.894	0.610	−0.856
ToC	0.943	0.968	0.964	0.985	0.980	1.000	0.956	0.993	0.988	0.429	−0.981	0.989	0.864	0.990	0.983	−0.986	−0.977	1.000	0.970	1	0.759	0.785	−0.956
TSN	0.932	0.899	0.904	0.636	0.874	0.770	0.916	0.678	0.849	0.914	−0.619	0.654	0.983	0.658	0.866	−0.858	−0.602	0.760	0.894	0.759	1	0.192	−0.534
Avail. K	0.534	0.603	0.593	0.879	0.644	0.774	0.569	0.852	0.681	−0.223	−0.889	0.868	0.367	0.866	0.657	−0.669	−0.899	0.783	0.610	0.785	0.192	1	−0.932
EC	−0.804	−0.851	−0.844	−0.992	−0.877	−0.951	−0.828	−0.984	−0.900	−0.145	0.995	−0.989	−0.678	−0.988	−0.886	0.892	0.997	−0.955	−0.856	−0.956	−0.534	−0.932	1

Table 8.
Correlation coefficient among the various variables used for nematode community analysis.

6. Evenness indices

The values for both the evenness indices (Sheldon and Heip indices) were observed to be the same for the two Zones 'A' and 'B' and calculated to be 0.82 ± 0.00 (0.82–0.82) and 0.72 ± 0.00 (0.72–0.72), respectively. For Zone 'C' the values were inestimable (**Table** 7).

7. Maturity index

In Zone 'A' where there was the least disturbance, the MI was highest and a mean value of 3.46 ± 0.26 (3.36–3.87) was observed representing the structured soil ecosystem. Zone 'B' which represented coal mine-managed spoil of 21 years period, showed an MI equal to 2.96 ± 0.50 (2.87–3.12) while in Zone 'C' MI was incalculable (**Table** 7 and **Figure** 7).

8. Plant parasitic index (PPI)

In Zone 'A' the PPI was calculated to be 0.78 ± 0.09 (0.50–1.25) whereas in Zone 'B' the PPI value was 3.26 ± 0.23 (2.00–3.65). The PPI and MI showed an inverse correlation (**Table** 7 and **Figure 8**).

9. Trophic diversity index

The value of trophic diversity index of Zone 'A' was 1.17 ± 0.02 (1.06–1.32), while those of Zone 'B' and Zone 'C' were 1.09 ± 0.04 (1.04–1.25) and 1.00 ± 0.00 (1.04–1.25), respectively (**Table 8** and **Figure 8**).

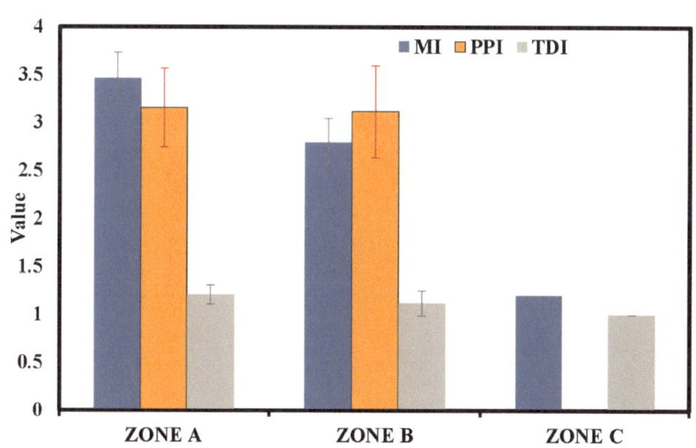

Figure 8.
Comparison of MI, PPI, and TDI in three different zones of coal mine region.

10. Weighted faunal analysis

The indices calculated depended on weighted nematode indicator guilds [1, 12] to assess the level of organic enrichment (EI), the decomposition pathways (CI) (**Table 8** and **Figure 9**), the basal/minimal level of resource utilization (BI) and the structured status or stability of ecosystem (SI). In Zone 'A' the structure index (SI) was highest i.e., 68.23 ± 10.20 (58.30–77.43), while in Zone 'B' the SI value was calculated to be 45.15 ± 5.85 (40.68–52.68) and in Zone 'C' the SI was zero. On the contrary, the enrichment index (EI) was calculated to be 12.62 ± 4.66 (7.65–32.25), 22.62 ± 12.58 (16.53–64.65) and 9.54 ± 1.54 (9.50–10.56) in Zone 'A', Zone 'B' and Zone 'C', respectively. The values of basal index (BI) for three different Zones were 74.46 ± 3.25 (70.13–78.16) in Zone 'C' followed by value of 57.29 ± 8.10 (44.24–67.35) in Zone 'B'; lowest value of BI was observed in 41.28 ± 12.30 (32.04–62.05) in Zone 'A'. The values of channel index (CI) in Zone 'A' and Zone 'B' were 0.31 ± 0.40 (0.25–0.84) and 0.54 ± 0.42 (0.50–1.24), respectively while CI was incalculable in Zone 'C' (**Table 8** and **Figure 10**).

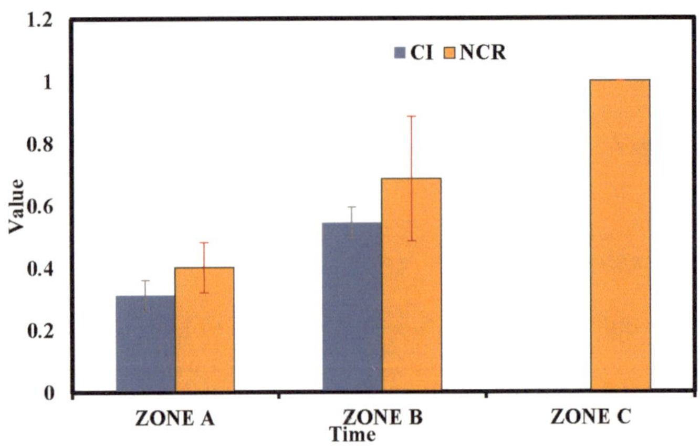

Figure 9.
Comparison of CI and NCR in three different zones of coal mine region.

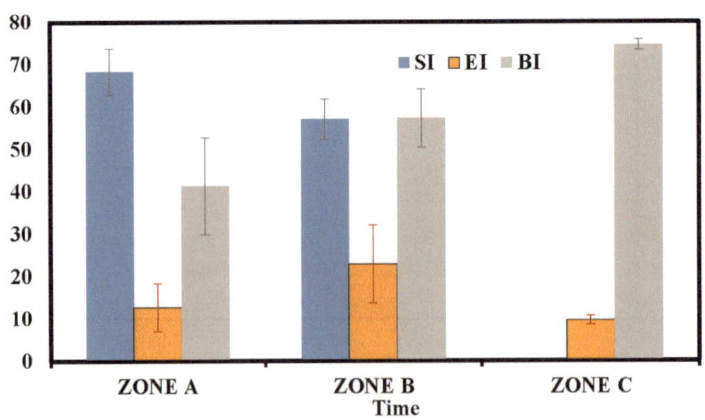

Figure 10.
Comparison of SI, EI, and BI in three different zones of coal mine region.

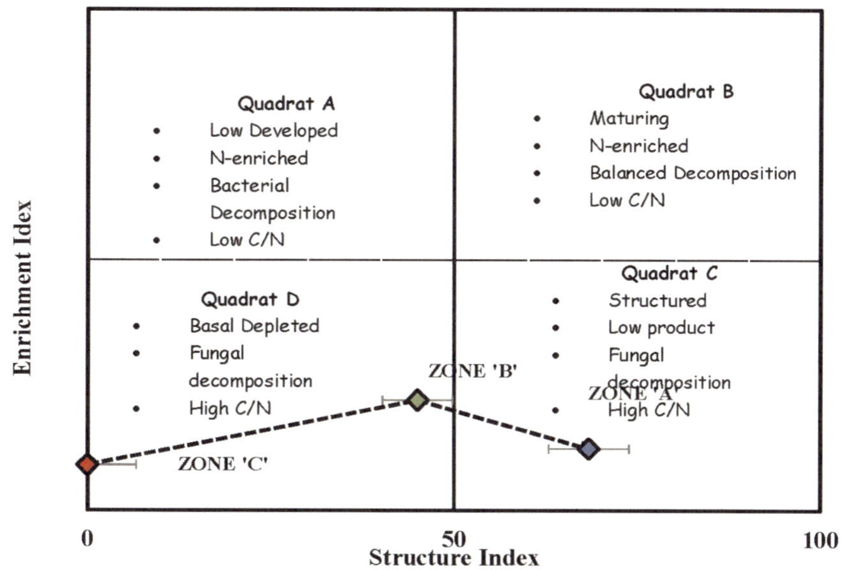

Figure 11.
Faunal profiles representing the food-web condition in relation to its structure (SI) and enrichment (EI) as indicated by the "weighted faunal analysis" (mean ± SE).

In faunal ordination graph between structure index and enrichment index, it was observed that in Zone 'A' the mean values were placed in quadrat 'C' whereas the mean values for Zone 'B' were at the interjunction of quadrat 'A' and quadrat 'C'. In Zone 'C' the soil had almost no vegetation, low nutrient, and very little organic carbon (**Figure 11**).

11. Correlations among different variables

The Principal component analysis (PCA) was carried out and a correlation circle was retrieved (**Figure 12**), which depicted the correlations among various variables. Structure index (SI) showed a positive correlation with number of bacterivores species ($r = 0.98$, $p < 0.05$), fungivores ($r = 0.99$, $p < 0.05$), and herbivores ($r = 0.99$, $p < 0.05$), omnivores ($r = 0.94$, <0.05) and predators ($r = 0.99$, $p < 0.05$). Almost all the trophic groups showed a positive relationship with SI. Structure index also showed a positive correlation with PPI ($r = 0.99$, $p < 0.05$), pH ($r = 1.00$, $p < 0.05$), TDI ($r = 0.96$, $p < 0.05$), NCR ($r = 0.93$, $p < 0.05$) and MI ($r = 0.99$, $p < 0.05$). SI also depicted low positive relationship EI ($r = 0.56$, $p < 0.05$) and strong negative correlation with CI ($r = 0.99$, $p < 0.05$) and BI ($r = -0.94$, $p < 0.05$). EI showed strong negative correlation with CI and BI all at ($p < 0.05$). All trophic groups showed a positive correlation among themselves ($p < 0.05$) and also with PPI, pH, and MI. Among the trophic groups, only bacterivores showed robust positive correlation with EI ($r = 0.70$, $p < 0.05$). EI did not show any correlation with PPI, nonsignificant correlation with SI and significant positive correlation with MI. Soil organic carbon was significantly positively correlated with all trophic groups ($p < 0.05$) and show a negative correlation with electrical conductivity, porosity, basal index and bulk density (**Table 8** and **Figure 13**).

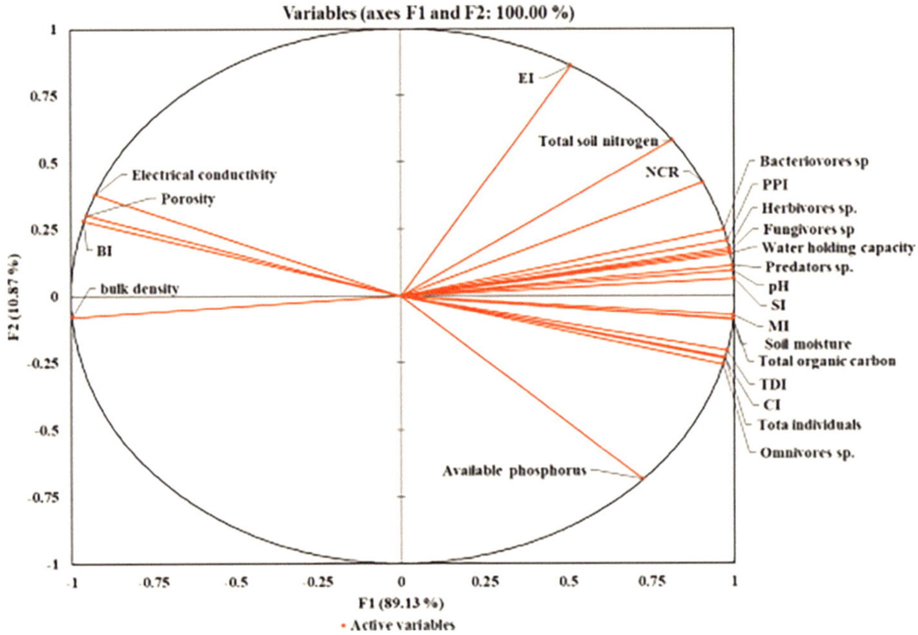

Figure 12.
Biplot correlation circle depicting the correlation among the different variables.

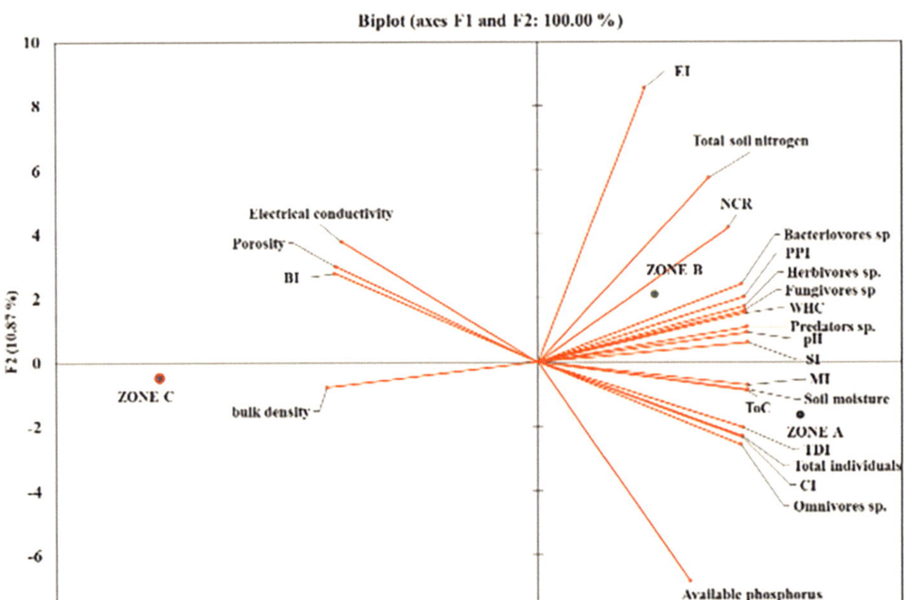

Figure 13.
Biplot ordination (PCA) of observation dates on trophic groups abundance in terms of species, total nematode abundance and various other variables with respect three different zones in coal mine region.

In Biplot ordination graph the three Zones can be clearly differentiated with respect to biotic and abiotic variables and can be superimposed on the anthropogenic disturbances to assess its impact.

12. Discussion

Human activities disturb the soil ecosystem and affect soil nematode diversity [8, 23–25]. The effect is well reflected in present study where there is a significant impact on below-ground soil biodiversity with special reference to nematodes. The transition from aboveground plant heterogeneity in natural forests to extreme loss of vegetation due to mining activities is reflected in the changed community characteristics of soil nematodes and in their species diversity.

In Zone 'A', the number of species (species richness) was greater, but species dominance was low while in Zone 'B', an intermediate value of species richness with high species dominance of bacterial-feeders was recorded. These species *viz.*, *Achromadora porosus* sp. n., *Acrobeloides nanus* and *Acrobeles cylindricus* showed tolerance to environmental stress, seasonal anhydrobiosis and other anthropogenic disturbances. Further, Zone 'B' represented the reclaimed 21 years old coal mine spoil with some incorporation of amendments to enrich the soil thus another reason for abundance of bacterivores. In Zone 'C' only three species were found *viz.*, *Mesorhabditis tenuhystera* sp. n., *Mesorhabditis vernalis* and *Acrobeloides nanus*. As, Zone 'C' was the active site for mining and with constant disturbances, the soil was almost lifeless with no nutrients and the area simply represented old exposed rocks mixed with coal dust having little organic carbon. The chemical residue from blasting and heavy metal contamination from mining process hindered the growth of plants and animals. The rhabditid species (*cp-1*) found in active mining Zone 'C', were enrichment opportunists and might have passively reached the active mining site *via* wind, water or through phoresis [26]. However, their sustenance showed their greater tolerance abilities towards mineral and industrial waste materials [27]. Thus *Mesorhabditis* spp. demonstrated greater endurance. In unfavorable periods, production of dauer larvae was the possible survival strategy for these enrichment opportunists. Zone 'B', the relatively less disturbed site, might have initially shown the growth of zymogenous (fast growing) microflora on fossil organic carbon that formed food for the microbivorous species, a condition also noticed by Háněl [28] who reported the abundant populations of genera *Acrobeloides* and *Aphelenchoides*. Zone 'A' consisting of natural wild forest, showed heterogenetic topography, and the litter accumulation due to undulating surfaces, was the reason for relatively high diversity and abundance of all trophic groups including bacterivores with values of both Simpson's index and Shannon's index comparable to any natural forest [23, 29–32].

Although there was not a major difference between the total number of species of fungivores and herbivores in Zone 'A' and Zone 'B', the mean abundance of both the trophic groups was much high in Zone 'A'. Despite the fact that stable ecosystems like Zone 'A' demonstrated species diversity and evenness, very often the species diversity has been reported to be greater in habitats subject to intermediate levels of disturbance because stochastic, intermediate (partial) elimination of resources by disturbance leads to species-specific mortality allowing the co-existence of competitively inferior species [33]. The combination of intermediate level of disturbance with intermediate productivity levels demonstrated a peak in species richness, not only due to periodic decreases of competitively dominant species but also to increased niche packing [34]. Structurally, complex environments thus provide more niches thereby increasing species diversity [35]. In conformity with the earlier reports on increased diversity at intermediate disturbance [36–38] and a positive relationship between habitat complexity and species diversity [39–44], Zone B demonstrated similar status. It showed an intermediate level of disturbances and the species diversity too was somewhat closer to Zone A which represented undisturbed zone. Large mononchid and large dorylaimoid nematodes with higher *c-p* values were mostly missing from Zone 'B' (**Table 5**), except for *Discolaimus major* and *Discolaimoides bulbiferous* which constituted the predator group along with another low *c-p* value predator and disturbance or enrichment indicator *Mononchoides longicauda*. Zone 'C' simply represented bacterivore colonizers with *c-p* values 1 and 2. Thus the nematode community structure in the selected zones showed typical examples of maturity and species diversification with respect to anthropogenic disturbances.

The maturity index of Zone 'A' Zone 'B' did not show much difference and could well indicate the status of Zone 'A' as structured and mature type of ecosystem and Zone 'B' as moderately structured and maturing type. Zone 'C' was obviously a highly disturbed region because of mining activities. These results agree well with those reported by Háněl [28] on coal mines.

The faunal profile results also corresponded with MI values of Zone 'A' and Zone 'B' as both reflected the dominant fungal decomposition pathways whereas Zone 'C' with disturbed and degraded food web demonstrated bacterial decomposition channels.

Although the present study has provided some pertinent information related to nematodes found in coal mine areas and the differences in nematode communities and their gradual succession in different overburdens or spoils, a detailed discussion on the role of individual species could not be done due to space constraint. Nevertheless, it is clearly evident that nematode assemblages can be evaluated and can serve as excellent tools for environmental monitoring or environment quality assessment. The study has opened up avenues for more studies to be conducted in coal mine areas using nematodes as models. Also, future studies in this direction may further vouch for comparison and may indicate any specificity or association of nematode taxa to coal mine areas. Such information may also have predictive value about the specificity and occurrence of species in degraded habitats, early successional and late successional stages to undisturbed habitats and can be used in modeling and predicting future changes in biodiversity and species interactions with land use changes.

13. Conclusion

In conclusion, this study highlights the significant impact of human activities on soil nematode diversity, particularly in the context of mining activities. The transition from natural forests to mining sites results in notable changes in soil nematode communities and species diversity.

In Zone 'A', which represents undisturbed natural forests, there is a higher species richness and lower species dominance among nematodes. In Zone 'B', which is a reclaimed coal mine spoil with some soil enrichment measures, there is an intermediate level of species richness but a higher dominance of bacterial-feeders. These nematode species in Zone 'B' exhibit tolerance to environmental stress and anthropogenic disturbances. On the other hand, Zone 'C', an actively mined area with constant disturbances and little organic matter, contains only a few nematode species, primarily enrichment opportunists with a high tolerance for mineral and industrial waste materials.

The study also suggests that intermediate levels of disturbance, as seen in Zone 'B', can lead to increased species diversity due to periodic decreases of dominant species and increased niche packing. Habitat complexity, as observed in both Zone 'A' and Zone 'B', also contributes to higher species diversity.

The maturity index values of Zone 'A' and Zone 'B' indicate structured and mature ecosystems, while Zone 'C' represents a highly disturbed and degraded region due to mining activities. The faunal profile results align with the maturity index values, with Zone 'A' and Zone 'B' dominated by fungal decomposition pathways and Zone 'C' characterized by bacterial decomposition channels.

While this study provides valuable insights into nematode communities in coal mine areas, further research is needed to explore the roles of individual nematode species in these ecosystems. Nematodes can serve as excellent indicators for environmental monitoring and assessing environmental quality. Future studies in this area may reveal specific associations of nematode taxa with coal mine areas, aiding in biodiversity modeling and predicting how species interactions change with land use alterations. This research opens the door to a deeper understanding of the impact of human activities on soil ecosystems and the potential for ecological restoration efforts.

Author details

Mohammad Asif[1*] and P. Fazul Rahaman[2*]

1 Section of Zoology, Directorate of Distance Education, Maulana Azad National Urdu University Hyderabad, India

2 Department of Zoology, School of Sciences, Maulana Azad National Urdu University Hyderabad, India

*Address all correspondence to: ga2628@myamu.ac.in and rahmanf1965@gmail.com

References

[1] Neher DA, Williams KM, Lovell ST. Environmental indicators reflective of road design in a forested landscape. Ecosphere. 2017;**8**:e01734

[2] Yeates GW, Bongers T, De Goede RGM, Freckman DW, Georgieva SS. Feeding habits in soil nematode families and genera-an outline for soil ecologists. Journal of Nematology. 1993;**25**:315-331

[3] Costa SR, van der Putten WH, Kerry BR. Microbial ecology and nematode control in natural ecosystems. In: Davies K, Spiegel Y, editors. Biological Control of Plant-Parasitic Nematodes: Progress in Biological Control. Vol. 11. Dordrecht: Springer; 2011. pp. 39-64

[4] Ferris H, Matute MM. Structural and functional succession in the nematode fauna of a soil food web. Applied Soil Ecology. 2003;**23**:93-110

[5] Ferris H, Venette RC, Lau SS. Population energetics of bacterial feeding nematodes, stage specific development and fecundity rates. Soil Biology and Biochemistry. 1996;**28**: 271-280

[6] Ingham RE, Trofymow JA, Ingham ER, Coleman DC. Interactions of bacteria, fungi, and their nematode grazers: Effects on nutrient cycling and plant growth. Ecological Monographs. 1985;**55**:119-140

[7] Ferris H, Bongers T, De Goede RGM. Nematode faunal indicators of soil food web condition. Journal of Nematology. 1999;**31**:534-535

[8] Neher DA. Role of nematodes in soil health and their use as indicators. Journal of Nematology. 2001;**33**:161-168

[9] Ritz K, TrudgilL DL. Utility of nematode community analysis as an integrated measure of the functional state of the soils: Perspectives and challenges. Plant and Soil. 1999;**212**:1-11

[10] Bongers T. The maturity index: An ecological measure of an environmental disturbance based on nematode species composition. Oecologia. 1990;**83**:14-19

[11] De Goede RGM, Georgieva SS, Verschoor BC, Kamerman JW. Changes in nematode community structure in a primary succession of blown-out areas in a drift sand landscape. Fundamental and Applied Nematology. 1993;**16**:501-513. DOI: 10.1016/S0169-5347(98)01583-3

[12] Ferris H, Bongers T, De Goede RGM. A framework for soil food web diagnostics: Extension of nematode faunal analysis concept. Applied Soil Ecology. 2001;**18**:13-29

[13] Hu J, Chen G, Hassan WM, Chen H, Li J, Du G. Fertilization influences the nematode community through changing the plant community in the Tibetan Plateau. European Journal of Soil Biology. 2017;**78**:7-16

[14] Kudrin A. Effects of low quantities of added labile carbon on soil nematodes in intact forest soil microcosms. European Journal of Soil Biology. 2017; **78**:29-37. DOI: 10.1016/j.ejsobi. 2016.11.002

[15] Neher DA, Darby BJ. Computation and application of nematode community indices: General guidelines. In: Eyualem A, Traunspurger W, Andrassy I, editors. Freshwater Nematodes: Taxonomy and Ecology. Wallingford, UK: CAB International; 2006. pp. 211-222

[16] Nethi S, Grewal PS, de Nardo EA, Stinner BR. Non-target effect of entomopathogenic nematodes on soil nematode community. Applied Soil Ecology. 2002;**39**:735-744

[17] Schorpp Q, Schrader S. Dynamic of nematode communities in energy cropping systems. European Journal of Soil Biology. 2017;**78**:92-101. DOI: 10.1016/j.ejsobi.2016.12.002

[18] Sohlenius B. Influence of cropping system and nitrogen input on soil fauna and microorganisms in a Swedish arable soil. Biology and Fertility of Soils. 1990; **9**:168-173

[19] Yeates GW, Bongers T. Nematode diversity in agroecosystems. Agriculture Ecosystems and Environment. 1999;**74**: 113-135

[20] Zhao J, Li D, Fu S, He X, Fu Z, Zhang W, et al. Using the biomasses of soil nematode taxa as weighting factors for assessing soil food web conditions. Ecological Indicators. 2016;**60**:310-316. DOI: 10.1016/j.ecolind.2015.06.003

[21] Neher DA, Wu J, Barbercheck ME, Anas O. Ecosystem type affects interpretation of soil nematode community measures. Applied Soil Ecology. 2005;**30**:47-64

[22] Yeates GW. Soil nematodes in terrestrial ecosystems. Journal of Nematology. 1979;**11**:117-212

[23] Freckman DW, Ettema CH. Assessing nematode communities in agroecosystem of varying human intervention. Agriculture, Ecosystems and Environment. 1993;**45**:239-261

[24] Ito T, Araki M, Higashi T, Komatsuzaki M, Kaneko N, Ohta H. Responses of soil nematode community structure to soil carbon changes due to different tillage and cover crop management practices over a nine-year period in Kanto, Japan. Applied Soil Ecology. 2015;**89**:50-58. DOI: 10.1016/j.apsoil.2014.12.010

[25] Zhao J, Neher D. Soil nematode genera that predict specific types of disturbance Appl. Soil Ecology. 2013;**64**: 135-141

[26] Rehfeld K, Sudhaus W. Vergleichende Untersuchung des Sexualverhaltens eines Zwillingsarten-Paares von Rhabditis (Nematoda) – Etude comparée du comportement sexuel de deux espèces sœurs de Rhabditis (Nematoda) – Comparative study of sexual behaviour of two sibling species of Rhabditis (Nematoda) Zoologische Jahrbucher. Abteilung für Systematik, Okologie und Geographie der Tiere. 1985;**112**(4):435-454

[27] Dmowska E. Nematodes colonizing power plant ash dumps. I. Soil nematodes in ash dumps non-reclaimed and reclaimed by adding mineral soil and sowing grass. Polish Journal of Ecology. 2001;**49**:231-241

[28] Háněl L. Succession of soil nematodes in pine forests on coal-mining sands near Cottbus, Germany. Applied Soil Ecology. 2001;**16**:23-34

[29] Yeates GW, King KL. Soil nematodes as indicators of the effect of management on grasslands in the New England Tablelands (NSW): Comparison of native and improved grasslands. Pedobiologia. 1997;**41**:526-536

[30] Pen-Mouratov S, Hu C, Hindin E, Steinberger Y. Soil microbial activity and a free-living nematode community in the playa and in the sandy biological crust of the Negev Desert. Biology and Fertility of Soils. 2011;**47**:363-375. DOI: 10.1007/s00374-011-0540-x

[31] Liang W, Lavian I, Steinberger Y. Dynamics of nematode community composition in a potato field. Pedobiologia. 1999;**43**:459-469

[32] Liang W, Lavian I, Steinberger Y. Effect of agriculture management on nematode communities in a Mediterranean agroecosystem. Journal of Nematology. 2001;**33**:208-213

[33] Huston M. A general hypothesis of species diversity. American Naturalist. 1979;**113**:81-101. DOI: 10.1086/283366

[34] Huston MA. Biological Diversity: The Coexistence of Species on Changing Landscapes. Cambridge: Cambridge University Press; 1994. 681 pp

[35] MacArthur R, Wilson OE. The Theory of Island Biogeography. Princeton: Princeton University Press; 1967

[36] Aronson RB, Precht WF. Landscape patterns of reef coral diversity: A test of the intermediate disturbance hypothesis. Journal of Experimental Marine Biology and Ecology. 1995;**192**:1-14. DOI: 10.1016/0022-0981(95)00052-S

[37] Flöder S, Sommer U. Diversity in planktonic communities: An experimental test of the intermediate disturbance hypothesis. Limnology and Oceanography. 1999;**44**:1114-1119. DOI: 10.4319/lo.1999.44.4.1114

[38] Huxham M, Roberts I, Bremner J. A field test of the intermediate disturbance hypothesis in the soft-bottom intertidal. International Review of Hydrobiology. 2000;**85**:379-394. DOI: 10.1002/1522-2632(200008)85:4<379::AID-IROH379>3.0.CO;2-X

[39] Cramer MJ, Willig MR. Habitat heterogeneity, species diversity and null models. Oikos. 2005;**108**:209-218

[40] Davidowitz G, Rosenzweig ML. The latitudinal gradient of species diversity among North American grasshoppers (Acrididae) within a single habitat: A test of the spatial heterogeneity hypothesis. Journal of Biogeography. 1998;**25**:553-560

[41] French DD, Picozzi N. 'Functional Groups' of bird species, biodiversity and landscapes in Scotland. Journal of Biogeography. 2002;**29**:231-259. DOI: 10.1046/j.1365-2699.2002.00664.x

[42] Hendrickx F, Maelfait JP, Van Wingerden W, Schweiger O, Speelmans M, Aviron S, et al. How landscape structure, land-use intensity and habitat diversity affect components of total arthropod diversity in agricultural landscapes. Journal of Applied Soil Ecology. 2007;**44**:340-351. DOI: 10.1111/j.1365-2664.2006.01270.x

[43] Kerr JT, Packer L. Habitat heterogeneity as a determinant of mammal species richness in high-energy regions. Nature. 1997;**385**:252-254. DOI: 10.1038/385252a0

[44] Tews J, Brose U, Grimm V, Tielborger K, Wichmann MC, Schwager M, et al. Animal species diversity driven by habitat heterogeneity/diversity: The importance of keystone structures. Journal of Biogeography. 2004;**31**:79-92

Chapter 2

Ecology, Adaptation, and Parasitism of Entomopathogenic Nematodes

Lalson Wesly Johnson, Rajaswaminathan Vairavan, Venkadesh Ganesan, Gurram Mallikarjun and Katakam Rupini Krishna

Abstract

Entomopathogenic nematodes (EPNs) are a distinct group of insect parasitic nematodes widely used in biological pest control. Nematodes in Steinernematidae and Heterorhabditidae have a mutual association with pathogenic bacteria of Enterobacteriaceae family to kill insect hosts rapidly. In this book chapter, we would like to address the effect of ecology, behavior, symbiosis, and parasitism of EPNs for their entomopathogenic potential under field conditions in positive and negative way. Hence, this chapter will focus on four objectives—(1) The impact of biotic and abiotic factors in abundance, dispersal and persistence of EPNs, (2) the finding behavior of EPNs, (3) EPN adaptation strategies for survival during stress conditions, and (4) nature of nematode-bacterium symbiotic relationship and their role in killing insect pests. Through a comprehensive literature review and analysis, this chapter will contribute much to the existing knowledge on EPNs, emphasizing their ecological significance and the potential implications for sustainable pest control practices.

Keywords: entomopathogenic nematodes, ecology, adaptation, parasitism, symbiosis, *Heterorhabditis*, *Steinernema*, *Photorhabdus*, *Xenorhabdus*, biological control

1. Introduction

In the intricate tapestry of Earth's ecosystems, countless organisms engage in a relentless dance for survival and dominance. Among these, entomopathogenic nematodes, or nematode parasites, stand as fascinating actors in the theater of life. These diminutive yet potent creatures have evolved remarkable strategies to navigate their environments, exploit their host insects, and perpetuate their lineage. The chapter before you embarks on a journey into the captivating world of entomopathogenic nematodes, shedding light on their ecology, intricate adaptations, and their sinister yet awe-inspiring parasitic lifestyle.

From the labyrinthine soil ecosystems to the leafy canopies of towering forests, entomopathogenic nematodes have carved out niches as nature's clandestine assassins. Among the diverse insect-parasitic nematodes, entomopathogenic nematodes (EPNs) are distinct and cooperate with insect-pathogenic symbiont bacteria to kill insect hosts. Nematodes in Steinernematidae and Heterorhabditidae have mutual association with pathogenic bacteria *Xenorhabdus* and *Photorhabdus*, respectively. EPNs kill insect hosts rapidly, usually within 48 hours of infection. Hence, they are being used worldwide for the biological management of insect pests of crops [1]. Their intricate interactions with their insect hosts, encompassing an array of nematode species and diverse host organisms, have confounded scientists for decades. In this chapter, we will unveil the enigmatic strategies these nematodes employ to locate, infect, and ultimately consume their unsuspecting prey.

The story of entomopathogenic nematodes is a story of relentless adaptation. It is a tale of microscopic creatures that have evolved an array of biological weaponry, from lethal mouthparts to symbiotic bacteria, enabling them to thrive in a world rife with challenges. Our exploration will delve into these adaptive marvels, providing insights into the evolutionary arms race that has fueled their success as formidable parasites. Yet, as we delve deeper into the world of entomopathogenic nematodes, we will also uncover the paradoxical intricacies of parasitism. While they spell doom for their insect hosts, these nematodes form intricate relationships with other organisms, particularly symbiotic bacteria, which assist them in their gruesome task. The chapter will unravel the symbiotic partnerships that sustain their parasitic lifestyle and illuminate the complex web of life in which they are entangled.

As we embark on this journey through the ecology, adaptation, and parasitism of entomopathogenic nematodes, we invite you to peer through the microscope and into the miniature worlds they inhabit. Prepare to be captivated by their astonishing strategies, awed by their adaptations, and intrigued by the profound ecological implications of their parasitic existence. These small but remarkable nematodes have much to teach us about the intricate balance of life on our planet.

1.1 Internal stress

During their free-living stage, entomopathogenic nematodes (EPNs) encounter a range of internal and environmental stressors. The capacity to effectively navigate these challenges assumes paramount importance in determining their success in infection and subsequent survival [2]. Various attributes of EPNs come into play, influencing their resilience in the face of both internal and external stressors, thereby shaping their survival as infective juveniles (IJs) within the soil.

1.1.1 Oxidative stress

Oxidative stress occurs when there is a disparity in the generation of reactive species (RS), including reactive oxygen species (ROS) and reactive carbonyl species (RCS), resulting from mitochondrial respiration and the antioxidant mechanisms of the nematode and its host. Due to their high reactivity, RS can interfere with cellular metabolism, and an overabundance or imbalance in their production can potentially lead to cellular demise [3]. To mitigate the effects of RS, diverse mechanisms are employed for detoxification, encompassing processes such as oxidation (e.g., facilitated by aldehyde dehydrogenases), conjugation (e.g., through interaction with glutathione), or reduction (e.g., facilitated by aldoketoreductases) [4]. In *Heterorhabditis bacteriophora*, IJ tolerance to H_2O_2 exposure correlates with their lifespan in sand [5]. Nematodes produce numerous enzymes dedicated to neutralizing reactive oxygen species (ROS) throughout their life cycle. While the effects of ROS and their detoxification mechanisms on *Caenorhabditis elegans* are well-documented, the influence of oxidative stress on the endurance of EPN infective juveniles (IJs) in soil is still not fully understood and warrants further exploration [6].

1.1.2 Nutritional stress

Infective juveniles (IJs) are the non-feeding stage and instead rely on their internal stored reserves of lipids, mainly triacylglycerols, and glycogen for their survival. These lipids make up roughly 20 to 31% of the nematode's dry weight. In *Heterorhabditis bacteriophora,* the amount of unsaturated fatty acids accounts for 57% of the total fatty acids. But Steinernematid species tend to have a higher amount of saturated fatty acids, at times up to 70% of their total lipid content.

Among the 18 fatty acids detected in *Steinernema* species, oleic acid, stearic acid, and palmitic acid decrease over time, showing that these fatty acids are used as primary energy sources by the nematodes. On the contrary, glycogen reserves range from 10 to 18% and they appear to deplete more slowly than lipids in early stage of IJs but more quickly following lipid depletion. As a result, it is proposed that glycogen might serving as an alternate energy source if lipid reserves in older IJs are exhausted [7].

2. Abiotic stress

2.1 Temperature

In temperate regions, Arctic, sub-Arctic, and high-altitude environments, nematodes that live in soil may experience temperatures below zero [8]. These organisms can persist in these circumstances by either tolerating the ice that forms on the exterior of their bodies (freezing tolerance) or by supercooling to prevent freezing at subzero temperatures (freezing avoidance). Nematodes exercising freeze avoidance technique perish when their bodies freeze. Supercooling is the capacity of an organism to maintain its body fluids in a liquid condition even at the temperatures below the freezing point. This ability can be enhanced by the production of the cryoprotectants such as trehalose and glycerol [8].

Entomopathogenic nematodes have been observed in polar regions, signifying their capacity to endure subzero temperatures. Steinernematids have been identified in multiple locales across northern Europe and Canada, characterized by their exposure to freezing climatic conditions. While Heterorhabditids have been isolated in temperate regions, the mechanisms underlying their cold tolerance remain unelucidated. Notably, *Steinernema feltiae*, *S. anomali*, and *Heterorhabditis bacteriophora* exhibit resistance to freezing, with recorded lower lethal temperatures of −22, −14, and −19°C, respectively [9].

It was also found that entomopathogenic nematodes can be cryopreserved using liquid nitrogen. The strains used nowadays for commercial applications are active at temperatures of 18 to 30 degrees [10]. However, some of the studies have shown that the new isolates from temperate regions can also effectively kill insects at temperatures ranging from 6 to 12°C [11].

2.2 Heat tolerance

Extreme temperatures, exceeding 32°C, have detrimental effects on various living organisms, including nematodes [8]. Heat shock proteins are stress-related proteins that play an important role in the survival of organisms at elevated temperatures.

Genes related to the production of these heat shock proteins were highly conserved across various species of EPNs and other free-living nematodes like *C. elegans*. The most commonly studied HSP that has been detected is HP88 in *H. bacteriophora* [12].

2.3 Desiccation tolerance

Besides extreme temperatures, dry conditions can also have a negative impact on the survival of the nematodes [13]. For movement and efficient survival of nematodes, they require at least a thin film of water surrounding the body. But some of the stages in their life cycle can withstand the absence of water for a prolonged period. This can be possible by the process called anhydrobiosis, a type of cryptobiosis that is induced by prolonged exposure to dry conditions. Anhydrobiosis is achieved by the gradual loss of body water [14]. Besides, these nematodes also show some structural and behavioral changes such as clumping and coiling to reduce the exposed cuticular surface area, thus reduced the loss of water through the cuticle. While the specific biochemical pathways of anhydrobiosis are unknown, it has been found that during a progressive water loss process, the levels of trehalose and glycerol grow dramatically, while glycogen and lipid levels are dropped. Glycerol and trehalose act as the protectants [13].

In cases of severe dehydration, entomopathogenic nematodes (EPNs) undergo a process known as anhydrobiosis [15]. Anhydrobiosis is a reversible physiological phenomenon where an organism can lose as much as 98% of its body water without metabolic arrest, entering a state referred to as cryptobiosis. This concept was initially described by Cooper and van Gundy in 1971 and further elucidated by Womersley in 1981. While EPNs are capable of achieving partial anhydrobiosis, characterized by an approximate 80% reduction in oxygen consumption [16], they do not progress to full cryptobiosis. Therefore, they are categorized as quiescent anhydrobiotes [16–18]. Desiccation, the process of drying out, initially prompts a temporary increase in EPN metabolism before gradually slowing it down to levels below the normal metabolic rate [16]. During this phase, there is a decrease in glycogen production, indicated by the down-regulation of glycogen synthase (gsy-1), while the synthesis of trehalose and glycerol from existing glycogen and neutral lipid reserves increases. These findings have been supported by research from refs. [13, 15, 19–21]. Trehalose gradually replaces water in cell membranes and plays a crucial role in preserving cell structures and stabilizing proteins [14, 19, 20, 22]. In the case of *S. feltiae*, desiccation results in a twofold increase in trehalose content [23]. Furthermore, desiccation induces the expression of casein kinase (CK2), leading to the transcriptional activation of a nucleosome-assembly protein (NAP-1) through physical interaction [15, 24, 25]. In response to desiccation, EPNs also synthesize osmoregulant molecules (e.g., produced by ALDH) and antioxidants (e.g., Gg., DESC47, HSP40), which may serve to provide additional protection against damage caused by the drying process [15, 24, 25].

3. Biotic stress

3.1 Fungi

In fungi, nematophagous fungi are the common adversaries of nematodes, and they are classified under different categories such as endoparasites, nematode trapping, and parasites of cysts and eggs. If these fungi are present in the soil, this will be detrimental to the survival and ability of infection of EPNs [26]. In order to protect themselves from these fungi EPNs employ several mechanisms including, high cruising moment of nematodes to evade the traps formed by the fungi secondly some EPNs such as Heterorhabditids retain their second-stage cuticle until they enter the insect body, thus protecting their body from fungal traps [27]. Third, EPNs can detect the chemical signals from fungal traps and avoid those [28]. In addition, when the nematodes are encountered by the fungal traps, their defense system is activated leading to the production of antimicrobial peptides through PRRs, thereby ensuring the efficacy of EPN-based pest control strategies in the future.

3.2 Bacteria

Pathogenic soil bacteria can attach to the surface of the EPN, FLN, and plant parasitic nematodes (PPN) [29]. Main bacteria attacking these nematodes are *Pasteuria* sp., Kaistia sp., *Lysinibacillus fusiformis*, *Enterobacter* sp., *Bacillus cereus*, *Klebsiella quasipneumoniae*, and *Pseudomonas aeruginosa* [30]. Endospore-forming bacteria have been related to the reduction of virulence in EPNs. The immune response of EPNs to bacterial agents remains unclear. In the case of *C. elegans*, which is a model organism often used for studying nematode biology, pathogen attacks are detected through pattern recognition receptors (PRRs) that identify pathogen-associated molecular patterns and/or disruptions in cellular homeostasis. This detection triggers both cell-autonomous and non-autonomous responses. Understanding how EPNs identify and respond to various pathogens may give important insights about target markers that might improve EPN survival in soil [30].

4. Chemoreception behavior in entomopathogenic nematodes

Chemoreception in entomopathogenic nematodes plays a critical role in their survival, host-finding, and infection process. Chemoreception is the ability of organisms to detect and respond to chemical stimuli in their environment. EPNs use chemoreception to navigate through the soil in search of suitable insect hosts and to locate their hosts for infection. Here is how chemoreception works in entomopathogenic nematodes:

4.1 Host location

EPNs are free-living in the soil, and they actively search for suitable insect hosts to infect. To find hosts, they rely on chemical cues emitted by insects or decaying organic matter associated with the hosts. These chemical cues may include volatile compounds released by insect larvae, fecal matter, or other organic substances.

4.2 Movement toward host

Once the nematode detects the chemical cues associated with a potential host, it exhibits positive chemotaxis, which means it moves toward the source of the attractive chemical stimuli. This helps the nematode to approach the host effectively.

4.3 Recognition and penetration

When an EPN reaches an insect host, it needs to recognize and penetrate the insect's body to initiate infection. The nematode uses chemoreceptors located on its mouthparts to sense specific chemicals on the insect's surface, helping it to identify suitable entry points, such as natural body openings or softer cuticle areas.

4.4 Infection

After penetrating the host, the EPN releases symbiotic bacteria from its gut into the insect's body cavity. These bacteria are responsible for producing toxins that kill the insect and also help in breaking down the host tissues, creating a suitable environment for the nematodes to feed and reproduce.

Overall, chemoreception is a fundamental aspect of the biology of entomopathogenic nematodes as it allows them to locate, infect, and successfully parasitize their insect hosts. Understanding the chemoreception mechanisms in EPNs can aid in the development of more effective biocontrol strategies and improve their application in agricultural and pest management practices.

4.5 Foraging strategies

Infective juvenile (J3) is the only free-living stage of EPNs which are non-feeding, having thick cuticle, non-developing and non-reproductive stage called the dauer stage. There are two major types of foraging strategies in IJs, ambush (surprise attack by jumping on the host) and cruise (movement in the soil in search of host) [6]. The actively moving hosts are attacked by ambush foragers, while sedentary hosts are attacked by cruise foragers [31].

The EPN juveniles are attracted toward their host through volatile cues produced by host insect such as CO_2 [32, 33], excretory products and fecal matters [34], Heat [35], and pH gradient [36]. The EPN Juveniles are attracted to plant root diffusates also, because such plant is considered [10] as habitat for the host insect [37, 38]. For example, *Heterorhabditis megadis* which is attracted to plant diffusates when damaged by its host insect beetle *Diabrotica virgifera* [39] and by weevil larvae *Otiorhynchus sulcatus* [37].

5. GPCRs

G protein-coupled receptors (GPCRs) also called as seven transmembrane receptors which are located in the cell membrane of the nematode sensory structures. Signal transduction in G protein-coupled receptors (GPCRs) is a complex process by which extracellular signals are converted into intracellular responses. GPCRs are cell membrane receptors that transmit signals from various ligands, such as hormones, neurotransmitters, and sensory stimuli, to the inside of the cell [40]. Here is a general overview of the signal transduction process in GPCRs [41]:

5.1 Ligand binding

The process begins when a specific ligand binds to the extracellular domain of a GPCR. This binding induces a conformational change in the receptor, leading to activation.

5.2 G protein activation

Upon activation, the GPCR interacts with and activates a heterotrimeric G protein complex located on the intracellular side of the cell membrane. The G protein consists of three subunits: α, β, and γ.

5.3 G protein activation and dissociation

The binding of the activated GPCR to the G protein causes the exchange of GDP (guanosine diphosphate) on the α-subunit for GTP (guanosine triphosphate), leading to the dissociation of the α-subunit from the βγ-subunits.

5.4 Effector activation

Both the α-subunit and the βγ-subunits can independently regulate downstream signaling pathways. The α-subunit can directly interact with various effector proteins, such as adenylyl cyclase or phospholipase C, depending on the specific GPCR and the type of G protein involved.

5.5 Second messenger production

Activation of effector proteins leads to the production of second messengers, such as cAMP (cyclic adenosine monophosphate) or IP3 (inositol trisphosphate), depending on the pathway. These second messengers act as intracellular signaling molecules that mediate the transmission of the signal to various downstream effectors.

5.6 Activation of downstream pathways

The second messengers activate downstream signaling pathways that often involve protein kinases, ion channels, and other intracellular effectors. These pathways ultimately lead to changes in cellular responses, such as alterations in gene expression, enzyme activity, ion channel conductance, and more.

5.7 Termination of signal

The duration of the signaling response is tightly regulated to prevent continuous activation. The GTP-bound α-subunit of the G protein has intrinsic GTPase activity, which hydrolyzes GTP to GDP, resulting in the inactivation of the α-subunit. Once inactive, the α-subunit reassociates with the βγ-subunits to reform the inactive G protein complex.

5.8 Receptor desensitization and internalization

Prolonged or repeated stimulation of GPCRs can lead to desensitization, where the responsiveness of the receptor to ligand binding is reduced. This can involve processes such as phosphorylation of the receptor by kinases, leading to recruitment of β-arrestins that inhibit further G protein signaling. Additionally, the desensitized receptor can be internalized into the cell through endocytosis.

5.9 Receptor recycling or degradation

Following internalization, the receptor can be either recycled back to the cell membrane after ligand dissociation or targeted for degradation in lysosomes. The recycling and degradation processes help regulate the availability of active receptors on the cell surface.

Overall, GPCR signal transduction is a dynamic and highly regulated process that enables cells to respond to a diverse array of extracellular signals and adapt to changing environmental conditions. This process plays a pivotal role in various physiological and pathophysiological processes throughout the nematode body. These GPCRs will help in drug discovery and novel method of developing nematode resistance in plants for nematode management in case of PPNs and effective in host finding for beneficial nematodes like EPNs.

6. Insect parasitism and symbiosis

6.1 Parasitic potential of EPNs

After the application of EPNs in the field, their numbers decrease rapidly within the first few days, followed by a more gradual decline over the course of a month, influenced by several factors. Approximately 80 percent of the reduction in EPN populations post-application can be attributed to factors such as dehydration and exposure to UV radiation. Additionally, energy depletion caused by a lack of food resources and the presence of antagonists has a minor but significant negative impact on EPN survival once they are introduced into the field. For example, out of 1000 IJs that are initially applied, only around ten of them are capable of effectively overcoming these abiotic challenges and are available for insect biocontrol purposes. What makes EPNs unique is their ability to achieve successful insect pest control with a very small quantity of EPNs within 24–72 hours. Furthermore, this population can persist for years through recycling mechanisms [10].

6.2 Host penetration by EPNs

The one percent population of EPNs is also subjected to the insect's immune system as part of the process for achieving successful insect biocontrol. The primary role of the nematode is to effectively introduce symbiotic bacteria into the insect's hemocoel, which results in the successful elimination of insects through septicemia. The insect employs various defenses to counter the entry of the nematode, including behavioral, mechanical, and cellular responses. Typically, infective juveniles (IJs) initially enter the insect's body through natural openings such as the anus, mouth, or spiracles. However, there is an exception in the case of *Heterorhabditis*, which, while also using natural openings, can additionally pierce the insect's cuticle with their specialized tooth. This cuticle-piercing method allows them to gain entry rapidly, typically within 30 minutes to 1 hour, as opposed to the conventional route, which takes more than 3 hours to access the insect's hemocoel [2].

6.3 EPNs breaking the insect mechanical barriers

When insects detect the penetration of infective juveniles (IJs) through their cuticle, they exhibit aggressive behavioral defenses to avoid the nematodes. An example of this can be seen in white grubs, which employ grooming behavior using their rasters and legs, a behavior that can even result in the death of IJs [42]. Additionally, the epi-cuticular wall of mealy bugs has been observed to act as a barrier against Heterorhabditis nematodes. Furthermore, if the entry of EPNs through spiracles is detected by insects, they utilize sieve plates to protect their spiracles and cuticular hairs as mechanical resistance tools against IJ entry. For instance, white grubs, which possess long hairs, are capable of trapping IJs with these hairs [11]. Nevertheless, the mechanical pressure exerted by the IJs assists them in breaking through the relatively weak tracheolar wall of the insect [43]. Thirdly, although infective juveniles (IJs) prefer to enter the white grub through its mouth, caterpillars employ their sharp mandibles to kill these nematodes. Additionally, the passage of IJs into the mouth of wireworms is restricted due to the narrow width of their mouths. Therefore, in the case of most wireworms, houseflies, and leaf miners, IJs utilize the path provided through the anus for entry. However, this preference for entry through the anus can pose challenges for IJs, particularly when the insect frequently defecates. Once the nematodes successfully overcome these barriers, they enter the insect's gut. Nevertheless, within the foregut of the insect, gastric juices are produced, and these juices can result in the mortality of approximately 40 percent of the nematode population [44].

6.4 EPN shattering peritrophic membrane to gain entry into host midgut

After escaping from the foregut, the nematodes find themselves in the insect's midgut, which contains structures like gastric caecae, malphigian tubules, and midgut epithelium. This is where the discharge of waste materials with feces is terminated. At this stage, the nematodes will attempt to move deeper into the insect's hemocoel. However, their progress is often impeded by the insect's peritrophic membrane [43]. This peritrophic membrane serves to delay the penetration of nematodes into the insect's hemocoel. Within the midgut of the insect, epithelial cells secrete this chitinous envelope, which functions as an ultra-filter. Consequently, this layer restricts the passage of particles larger than 20 nanometers, effectively preventing nematodes from reaching the hemocoel. Despite the obstacle presented by the peritrophic membrane, *Heterorhabditis* and *Steinernema* can breach it by puncturing it using a mural tooth and creating holes through the application of mechanical pressure, given that the membrane is quite delicate. Additionally, hystolytic enzymes aid the nematode in breaking down and tearing the peritrophic membrane [45].

6.5 A look on nematode-bacteria synchronized life cycle

Once the infective juveniles (IJs) reach the insect's hemocoel, they undergo a complete recovery from the diapause stage. At this point, the nematodes feel secure and regurgitate their monospecific gut bacteria into the insect's hemocoel. The life cycle of these symbiotic bacteria becomes synchronized with that of their host. This is exemplified by *Heterorhabditis* and its monospecific symbiotic association with the *Photorhabdus* bacterium in their gut. Although host-restricted Photorhabdus bacteria are typically transmitted maternally in *Heterorhabditis* through endotokia matricida, the infective juveniles of these nematodes (J3), after recovering from diapause, regurgitate their entire gut bacteria into the insect's hemocoel as part of a selective mechanism (to distinguish between P form and M form bacterial cells). They then retrieve a portion of these bacteria, specifically the M form population, to form a persistent biofilm in their posterior intestinal cells, INT9L and INT9R [46]. As a result of both growth and the fresh adherence of bacteria, they form a mass consisting of more than 50 cells, firmly attaching themselves to the intestinal lumen on the posterior side of the nematode. Subsequently, these bacterial cells breach the gland epithelium and invade the cytoplasm of rectal gland cells. Each rectal gland cell of the maternal nematode contains at least one symbiont. These bacterial cells multiply within the rectal cells, resulting in around 10–30 vacuoles per cell, causing the rectal gland cells to enlarge. In parallel, the next generation of pre-infective juveniles (pre-IJs) develops inside the maternal nematode through a process known as endotokia matricida. These pre-IJs typically consume the protoplasm of the mother and develop within her body cavity. Meanwhile, the rectal gland cell undergoes lysis, and the vacuoles containing symbionts from the rectal cells enter the body cavity of the mother. Once the bacterial cells reach the maternal pseudocoelom, they attach themselves to the cardia of the freshly formed infective juveniles (IJs) and multiply in this location. Eventually, they reach the anterior intestine of the fresh IJs, which serves as their habitat for further reproduction and multiplication through the intestinal lumen [47, 48].

6.6 How nematode escapes encapsulation?

Conversely, you might be curious about the role of the P form population of bacterial cells, which will be discussed in the following paragraphs. Currently, both the nematode and bacteria are located in the midgut of the insect. However, it has been documented that it takes the nematode a minimum of 30 minutes to release the bacteria after entering the insect's hemocoel. In other words, the nematode has to contend with the insect's immune system for a brief period on its own. Once the infective juveniles (IJs) are detected within the insect's hemolymph, the insect's immune system initiates nematode encapsulation. This process involves the IJs becoming trapped within melanin-hardened cellular capsules, which have been observed in orthoptera, diptera, lepidoptera, and coleopteran insects [44]. Nematodes employ various strategies, including evasion, tolerance, and suppression, to escape from encapsulation by the insect's immune system. Firstly, infective juveniles (IJs) of *S. carpocapsae* secrete lipids that assist the nematode in evading recognition by the insect [49]. Secondly, at least a significant minority population of *Steinernema* nematodes can tolerate encapsulation and continue with their parasitic activities [50]. Lastly, sometimes, nematodes use a tactic of confusing the insect by regurgitating bacteria just before being encapsulated by the insect. This causes the insect's immune system to prepare for encapsulating the nematode, but the presence of bacteria goes unnoticed by the insect, allowing the bacteria to start killing hemocytes without interruption [51].

6.7 How bacteria respond to insect immune system?

After being recognized by the host's immune system, the bacteria undergo hemocyte aggregation and subsequent nodulation, processes partially triggered by eicosanoids. *Photorhabdus* and *Xenorhabdus* disrupt hemocyte aggregation and the formation of nodules by inhibiting the activity of phospholipase A2, an enzyme responsible for initiating the insect eicosanoid pathway. Additionally, *Photorhabdus*, similar to many other gram-negative pathogenic bacteria, possesses a type III secretion system (TTSS) that transfers effector proteins into the host's eukaryotic cells. One of these effectors, known as LopT, shields the cells by suppressing phagocytosis and diminishing nodulation. While *Xenorhabdus* lacks a dedicated TTSS, it relies on a flagellar TTSS for the secretion of lipase [52].

Lipopolysaccharide (LPS) is a significant component of the outer cell membrane found in both *Photorhabdus* and *Xenorhabdus*. In *Xenorhabdus*, LPS exhibits cytotoxic properties through its lipid A component, which binds to and damages insect hemocytes. Additionally, LPS has the ability to inhibit phenoloxidase activity, thereby suppressing melanization. In the case of *Photorhabdus*, the precise role of LPS is not entirely clear, but it may play a crucial role in countering antimicrobial peptides produced by the insect's immune system [53]. Conversely, *Xenorhabdus* takes a different approach by suppressing the transcription of insect genes responsible for encoding antimicrobial peptides. In the early stages of infection, *Photorhabdus* produces a trans stilbene antibiotic known as (E)-1,3-dihydroxy-2-(isopropyl)-5-(2-phenylethenyl) benzene. This compound serves a dual purpose in combating the host's immune response and providing protection against microbial competitors by inhibiting phenoloxidase activity [54].

Once they have successfully evaded the immune system, bacteria deploy various toxins to eliminate their insect hosts. Among the most potent weapons in their arsenal are the Tc (toxin complex) protein toxins. Tc proteins are high-molecular-weight insecticidal toxins that can cause lethal effects in insects, even when ingested orally. In the case of *Photorhabdus*, four distinct toxin complexes (Tca, Tcb, Tcc, and Tcd) have been identified. Notably, Tca shares similarities with *Bacillus thuringiensis*' d-endotoxin in its ability to disrupt the insect midgut epithelium [55]. In *Xenorhabdus*, the counterparts of Tc proteins are referred to as Xpt (*Xenorhabdus* protein toxin) and include XptA1, XptA2, XptB1, and XptC1. Among these, XptA1 plays a central role in exerting insecticidal activity, while XptB1 and XptC1 are crucial for full virulence. XptA1 possesses a unique "hollow box" structure that allows it to act as a receptacle for XptB1 and XptC1 proteins, enhancing its ability to bind effectively to the host gut. This cooperative action makes XptA1 a staggering 300 times more toxic to lepidopterous larvae when compared to its standalone potency. It is worth noting that the mode of action of Xpt toxins in *Xenorhabdus* differs from the toxin system employed by *Bacillus thuringiensis*. In other words, once bacteria bypass the host's immune defenses, they deploy Tc and Xpt protein toxins to target and kill their insect hosts. These toxins have evolved to be exceptionally potent, often disrupting the insect's gut and causing lethal effects, with XptA1 playing a central role in the process. Unlike similar toxins found in *Bacillus thuringiensis*, *Xenorhabdus*'s Xpt toxins employ a different mechanism of action [56].

Both of these bacterial genera release extracellular, cytotoxic proteins known as hemolysins. In *Xenorhabdus*, the xenorhabdolysin (C1) hemolysin stands out for its extremely high virulence, capable of triggering apoptosis in both insect and mammalian cells. This cytotoxin is a crucial component for *Xenorhabdus* to achieve full virulence. The role of hemolysins in *Photorhabdus* is somewhat less clear, as these bacteria can retain their virulence even without the secretion of hemolysins. Nonetheless, *Photorhabdus* produces other toxins that serve as significant virulence factors [54].

In *Photorhabdus*, the toxins Mcf1 and Mcf2 (makes caterpillars floppy) play a pivotal role in causing rapid loss of insect body turgor and ultimately death. Mcf1 accomplishes this by destroying hemocytes and the insect midgut through the induction of massive apoptosis. Meanwhile, the exact site and mode of action of Mcf2 remain unknown. These toxins share strong homology with each other but diverge at their N-termini, encoding different effector domains. These distinct effector domains may enable them to target different sites within the insect. Additionally, other *Photorhabdus* toxins are involved in damaging the insect gut. *Photorhabdus* takes up residence between the basal membrane and midgut epithelium and expresses the gut-active toxin Tca, along with PrtA, an RTX-like metalloprotease. Together, they induce extensive programmed cell death of the midgut epithelium [57].

Beyond causing the demise of the insect host, these toxins may aid in bioconverting the insect tissue to provide nourishment for the developing nematodes. Numerous virulence factors are engaged in the nematode-bacterial infection process. However, it remains unclear which mechanisms are universal and which are specific to nematodes, bacteria, or insect hosts. As new nematode species are being discovered, each with its unique association with specific bacteria, more opportunities arise to uncover the specificities involved in bacterial virulence.

7. Conclusion

Entomopathogenic nematodes exemplify the complexity of parasitic relationships in nature. Their remarkable adaptations and multifaceted parasitism strategies have far-reaching ecological implications, from influencing insect populations in natural ecosystems to providing sustainable solutions for pest management in agriculture. Understanding the ecology, adaptations, and parasitism mechanisms of EPNs is not only a subject of scientific interest but also holds promise for the development of eco-friendly pest control strategies.

Acknowledgements

We heartily thank the academic editor Soumalya Mukherjee, for providing us with this opportunity.

Conflict of interest

The authors declare no conflict of interest.

Author details

Lalson Wesly Johnson[1*], Rajaswaminathan Vairavan[2], Venkadesh Ganesan[1], Gurram Mallikarjun[1] and Katakam Rupini Krishna[1]

1 Division of Nematology, Indian Agricultural Research Institute, New Delhi, India

2 Department of Biology, Dalhousie University, Halifax, Nova Scotia, Canada

*Address all correspondence to: weslyjohnzen@gmail.com

References

[1] Dillman AR, Chaston JM, Adams BJ, Ciche TA, Goodrich-Blair H, Stock SP, et al. An entomopathogenic nematode by any other name. PLoS Pathogens. 2012;**8**(3):e1002527

[2] Shapiro-Ilan DI, Han R, Dolinksi C. Entomopathogenic nematode production and application technology. Journal of Nematology. 2012;**44**(2):206-217

[3] Sies H. On the history of oxidative stress: Concept and some aspects of current development. Current Opinion in Toxicology. 2018;7:122-126

[4] Detienne G, De Haes W, Mergan L, Edwards SL, Temmerman L, Van Bael S. Beyond ROS clearance: Peroxiredoxins in stress signaling and aging. Ageing Research Reviews. 2018;**44**:33-48

[5] Sumaya NH, Gohil R, Okolo C, Addis T, Doerfler V, Ehlers RU, et al. Applying inbreeding, hybridization and mutagenesis to improve oxidative stress tolerance and longevity of the entomopathogenic nematode *Heterorhabditis bacteriophora*. Journal of Invertebrate Pathology. 2018;**151**:50-58

[6] Huey RB, Pianka ER. Ecological consequences of foraging mode. Ecology. 1981;**62**(4):991-999

[7] Patel MN, Wright DJ. Fatty acid composition of neutral lipid energy reserves in infective juveniles of entomopathogenic nematodes. Comparative Biochemistry and Physiology Part B: Biochemistry and Molecular Biology. 1997;**118**(2):341-348

[8] Wharton D, Young S, Barrett J. Cold tolerance in nematodes. Journal of Comparative Physiology B. 1984;**154**:73-77

[9] Brown I, Gaugler R. Cold tolerance of steinernematid and heterorhabditid nematodes. Journal of Thermal Biology. 1996;**21**(2):115-121

[10] Glazer I. Survival and efficacy of *steinernema carpocapsae* in an exposed environment. Biocontrol Science and Technology. 1992;**2**(2):101-107

[11] Georgis R, Hague N. A neoaplectanid nematode in the larch sawfly Cephalcia lariciphila (hymenoptera: Pamphiliidae). Annals of Applied Biology. 1981;**99**(2):171-177

[12] Glazer I. Survival mechanisms of entomopathogenic nematodes. Biocontrol Science and Technology. 1996;**6**(3):373-378

[13] Womersley CZ, Gaugler R, Kaya H. Entomopathogenic nematodes in biological control. In: Dehydration Survival and Anhydrobiotic Potential. Boca Raton: CRC Press; 1990. p. 117

[14] Crowe JH, O'Dell SJ, Armstrong DA. Anhydrobiosis in nematodes: Permeability during rehydration. Journal of Experimental Zoology. 1979;**207**(3):431-437

[15] Gal TZ, Glazer I, Koltai H. Stressed worms: Responding to the post-genomics era. Molecular and Biochemical Parasitology. 2005;**143**(1):1-5

[16] Grewal P. Anhydrobiotic potential and long-term storage of entomopathogenic nematodes (Rhabditida: Steinernematidae). International Journal for Parasitology. 2000;**30**(9):995-1000

[17] Kaya HK, Gaugler R. Entomopathogenic nematodes. Annual

Review of Entomology. 1993;**38**(1): 181-206

[18] Simons WR, Poinar GO. The ability of Neoaplectana carpocapsae (steinernematidae: Nematodea) to survive extended periods of desiccation. Journal of Invertebrate Pathology. 1973;**22**(2):228-230

[19] Behm CA. The role of trehalose in the physiology of nematodes. International Journal for Parasitology. 1997;**27**(2):215-229

[20] Crowe J, Crowe L. Membrane integrity in anhydrobiotic organisms: Toward a mechanism for stabilizing dry cells. In: Somero GN, Osmond CB, Bolis CL, editors. Water and Life. Berlin, Heidelberg: Springer; 1992. pp. 87-103

[21] Qiu L, Bedding R. Energy metabolism and its relation to survival and infectivity of infective juveniles of Steinernema carpocapsae under aerobic conditions. Nematology. 2000;**2**(5):551-559

[22] Elbein AD, Pan Y, Pastuszak I, Carroll D. New insights on trehalose: A multifunctional molecule. Glycobiology. 2003;**13**(4):17R-27R

[23] Solomon A, Glazer I. Desiccation survival of the entomopathogenic nematode *Steinernema feltiae*: Induction of anhydrobiosis. Nematology. 1999;**1**(1):61-68

[24] Gal TZ, Glazer I, Koltai H. Differential gene expression during desiccation stress in the insect-killing nematode Steinernema feltiae IS-6. Journal of Parasitology. 2003;**89**(4):761-766

[25] Somvanshi VS, Koltai H, Glazer I. Expression of different desiccation-tolerance related genes in various species of entomopathogenic nematodes. Molecular and Biochemical Parasitology. 2008;**158**(1):65-71

[26] de Freitas Soares FE, Sufiate BL, de Queiroz JH. Nematophagous fungi: Far beyond the endoparasite, predator and ovicidal groups. Agriculture and Natural Resources. 2018;**52**(1):1-8

[27] Raja RK, Arun A, Touray M, Gulsen SH, Cimen H, Gulcu B, et al. Antagonists and defense mechanisms of entomopathogenic nematodes and their mutualistic bacteria. Biological Control. 2021;**152**:104452

[28] Willett DS, Alborn HT, Stelinski LL. Multitrophic effects of belowground parasitoid learning. Scientific Reports. 2017;**7**(1):2067

[29] Elhady A, Giné A, Topalovic O, Jacquiod S, Sørensen SJ, Sorribas FJ, et al. Microbiomes associated with infective stages of root-knot and lesion nematodes in soil. PLoS One. 2017;**12**(5):e0177145

[30] Loulou A, Mastore M, Caramella S, Bhat AH, Brivio MF, Machado RA, et al. Entomopathogenic potential of bacteria associated with soil-borne nematodes and insect immune responses to their infection. PLoS One. 2023;**18**(1):e0280675

[31] Campbell JF, Gaugler R. Nictation behaviour and its ecological implications in the host search strategies of entomopathogenic nematodes (Heterorhabditidae and Steinernematidae). Behaviour. 1993;**126**(3-4):155-169

[32] Lewis EE, Gaugler R, Harrison R. Response of cruiser and ambusher entomopathogenic nematodes (Steinernematidae) to host volatile cues. Canadian Journal of Zoology. 1993;**71**(4):765-769

[33] Gaugler R, Lebeck L, Nakagaki B, Boush GM. Orientation of the entomogenous nematode *Neoaplectana carpocapsae* to carbon dioxide 1. Environmental Entomology. 1980;**9**(5):649-652

[34] Schmidt J, All JN. Attraction of Neoaplectana carpocapsae (Nematoda: Steinernematidae) to common excretory products of insects. Environmental Entomology. 1979;**8**(1):55-61

[35] Byers JA, Poinar GO. Location of insect hosts by the nematode, Neoaplectana carpocapsae, in response to temperature. Behaviour. 1982;**79**(1):1-10

[36] Schmidt J, All JN. Chemical attraction of Neoaplectana Carpocapsae (Nematoda: Steinernematidae) to insect larvae. Environmental Entomology. 1978;7(4):605-607

[37] Van Tol RWHM, Van Der Sommen ATC, Boff MIC, Van Bezooijen J, Sabelis MW, Smits PH. Plants protect their roots by alerting the enemies of grubs. Ecology Letters. 2001;**4**(4):292-294

[38] Lei Z, Rutherford TA, Webster JM. Heterorhabditid behavior in the presence of the cabbage maggot, Delia radicum, and its host plants. Journal of Nematology. 1992;**24**(1):9-15

[39] Rasmann S, Köllner TG, Degenhardt J, Hiltpold I, Toepfer S, Kuhlmann U, et al. Recruitment of entomopathogenic nematodes by insect-damaged maize roots. Nature. 2005;**434**(7034):732-737

[40] Hart AC, Chao MY. From odors to behaviors in *Caenorhabditis elegans*. In: Menini A, editor. The Neurobiology of Olfaction [Internet]. Boca Raton (FL): CRC Press/Taylor & Francis; 2010. (Frontiers in Neuroscience). Available from: http://www.ncbi.nlm.nih.gov/books/NBK55983/ [Accessed: August 23, 2023]

[41] Liu N, Wang Y, Li T, Feng X. G-protein coupled receptors (GPCRs): Signaling pathways, characterization, and functions in insect physiology and toxicology. International Journal of Molecular Sciences. 2021;**22**(10):5260

[42] Gaugler R. Ecological Genetics of Entomopathogenic Nematodes. East Melbourne: CSIRO; 1993

[43] Forschler BT, Gardner WA. Parasitism of *Phyllophaga hirticula* (Coleoptera: Scarabaeidae) by *Heterorhabditis heliothidis* and *Steinernema carpocapsae*. Journal of Invertebrate Pathology. 1991;**58**(3):396-407

[44] Wang Y, Campbell JF, Gaugler R. Infection of entomopathogenic nematodes *Steinernema glaseri* and *Heterorhabditis bacteriophora* against *Popillia japonica* (Coleoptera: Scarabaeidae) larvae. Journal of Invertebrate Pathology. 1995;**66**(2):178-184

[45] Wang Y, Gaugler R. Host and penetration site location by entomopathogenic nematodes against Japanese beetle larvae. Journal of Invertebrate Pathology. 1998;**72**(3):313-318

[46] Somvanshi VS, Sloup RE, Crawford JM, Martin AR, Heidt AJ, Kim K suk, et al. A single promoter inversion switches *Photorhabdus* between pathogenic and mutualistic states. Science. 2012;**337**(6090):88-93

[47] Bhat CG, Budhwar R, Godwin J, Dillman AR, Rao U, Somvanshi VS. RNA-sequencing of heterorhabditis nematodes to identify factors involved in symbiosis with photorhabdus bacteria. BMC Genomics. 2022;**23**(1):741

[48] Ciche TA, Kim KS, Kaufmann-Daszczuk B, KCQ N, Hall DH. Cell invasion and matricide during *Photorhabdus luminescens* transmission by *Heterorhabditis bacteriophora* nematodes. Applied and Environmental Microbiology. 2008;**74**(8):2275-2287

[49] Dunphy GB, Webster JM. Partially characterized components of the epicuticle of dauer juvenile Steinernema feltiae and their influence on hemocyte activity in galleria mellonella. The Journal of Parasitology. 1987;**73**:584-588

[50] Thurston GS, Kaya HK, Gaugler R. Characterizing the enhanced susceptibility of milky disease-infected scarabaeid grubs to entomopathogenic nematodes. Biological Control. 1994;**4**(1):67-73

[51] Wang Y, Gaugler R. Steinernema glaseri surface coat protein suppresses the immune response of Popillia japonica (Coleoptera: Scarabaeidae) larvae. Biological Control. 1999;**14**(1):45-50

[52] Eleftherianos I, Heryanto C, Bassal T, Zhang W, Tettamanti G, Mohamed A. Haemocyte-mediated immunity in insects: Cells, processes and associated components in the fight against pathogens and parasites. Immunology. 2021;**164**(3):401-432

[53] Castillo JC, Reynolds SE, Eleftherianos I. Insect immune responses to nematode parasites. Trends in Parasitology. 2011;**27**(12):537-547

[54] Koppenhöfer HS, Gaugler R. Entomopathogenic nematode and bacteria mutualism. In: Defensive Mutualism in Microbial Symbiosis. Vol. 26. CRC Press; 2009. pp. 99-116

[55] Hinchliffe SJ, Hares MC, Dowling AJ. Insecticidal toxins from the photorhabdus and xenorhabdus bacteria. The Open Toxinology Journal. 2010;**3**(1):83-100

[56] Sheets JJ, Hey TD, Fencil KJ, Burton SL, Ni W, Lang AE, et al. Insecticidal toxin complex proteins from *Xenorhabdus nematophilus*. Journal of Biological Chemistry. 2011;**286**(26):22742-22749

[57] Dowling A, Waterfield NR. Insecticidal toxins from photorhabdus bacteria and their potential use in agriculture. Toxicon. 2007;**49**(4):436-451

Chapter 3

Plant-Parasitic Nematodes and Their Management: A Focus on New Nematicides

Arley Rey Páez

Abstract

Plant parasitic nematodes are microscopic organisms that inhabit soil and plant tissues. Among such organisms, those of the genera *Meloidogyne*, *Heterodera, Globodera,* and *Pratylenchus* spp., are the most harmful, as they affect more than 2500 different species of plants, generating worldwide economic losses of over US$100 billion per year. These nematodes constitute a notable threat to the country's progress and food security. Almost half of the global market for nematicides, which corresponds to US$ 1 billion per year, is used to control these nematodes. Non-fumigant nematicides are the most widely used in their control; however, many of them, such as carbamates and organophosphorus, are banned by environmental protection agencies because of their undesirable effects on non-target organisms. In the last 10 years, a new series of nematicides have emerged with different mechanisms of action than the old non-fumigant nematicides. Tioxazafen and fluazaindolizine are some of the latest new-generation nematicides that have come on the market. The rational design of new nematicides through *in silico* approaches combined with studies of the genetics and biochemistry of these microorganisms will help to better understand their management and control, aiming to reduce the environmental impact caused by the irrational use of nematicides.

Keywords: plant parasitic nematodes, root-knot nematodes, nematicide, tioxazafen, fluazaindolizine

1. Introduction

According to the United Nations (UN), the global population will reach 8.5 billion by 2030 [1]. As a result, the World Bank projects that food consumption will rise by 60–80% [2]. Although the Food and Agriculture Organization of the United Nations (FAO) has as one of its goals to end world poverty, the fact is that food security has been declining rapidly since 2015 [3]. Agricultural pests are clearly one of the many elements leading to a country's or region's food security being undermined [4]. Pests and plant diseases, according to the FAO, account for 40% of global food production losses, amounting to more than US$250 billion each year [5]. As a result, dealing with plant pests and diseases is critical to attaining higher agricultural sustainability and security, particularly in the post-pandemic age.

Although microorganisms play a vital role in agriculture, many of them act as phytopathogens, reducing agricultural yield seriously [6, 7]. Plant parasitic nematodes (PPNs) are microorganisms that have a deleterious impact on the harvested of numerous food crops such as soybean (*Glicine max*), corn (*Zea mays*), potato (*Solanum tuberosum*), tomato (*Solanum lycopersicum*), rice (*Oryza sativa*) and carrot (Daucus carota) [8, 9]. There are about 4000 PPNs species, but only a small number are linked to economic losses owing to decreased agricultural yields, which are estimated to be worth more than US$100 billion each year [10].

Because of their complex interaction with host plants, extensive host range, and the degree of damage caused by infection, root-knot nematodes (RKNs), cyst nematodes (CNs), and root-lesion nematodes (RLNs) were at the top of the list of the most economically important species [11]. Thus, *Heterodera glycines*, for example, has a significant influence on the agricultural economies of the United States of America (USA) and Brazil (**Figure 1**), since it affects one of the most important crops for global protein and carbohydrate production: soybeans [11]. Currently, Brazil leads the world in soybean production, with more than 114 million tons produced per year, followed by the USA with 97 million tons per year [12, 13]. In the USA alone, economic losses caused by *H. glycines* exceed US$1.0 billion per year [14].

Some of the approaches used to manage and control PPNs include resistant cultivars, crop rotation, biological management, and chemical compounds [8]. Chemical methods are the most efficient and therefore most widely used in the management of PPNs; however, many of the chemicals used are highly hazardous to human and environmental health, including carbamates (CMs), organophosphorus (OPs), organic halides, and pyridinylmethylbenzamides [15].

The arsenal of nematicides accessible for the management of PPNs has been substantially restricted by current environmental protection laws. This has increased demand for novel nematicides that are more selective, less toxic, and compatible with sustainable agricultural guidelines. As a result, various nematicides have

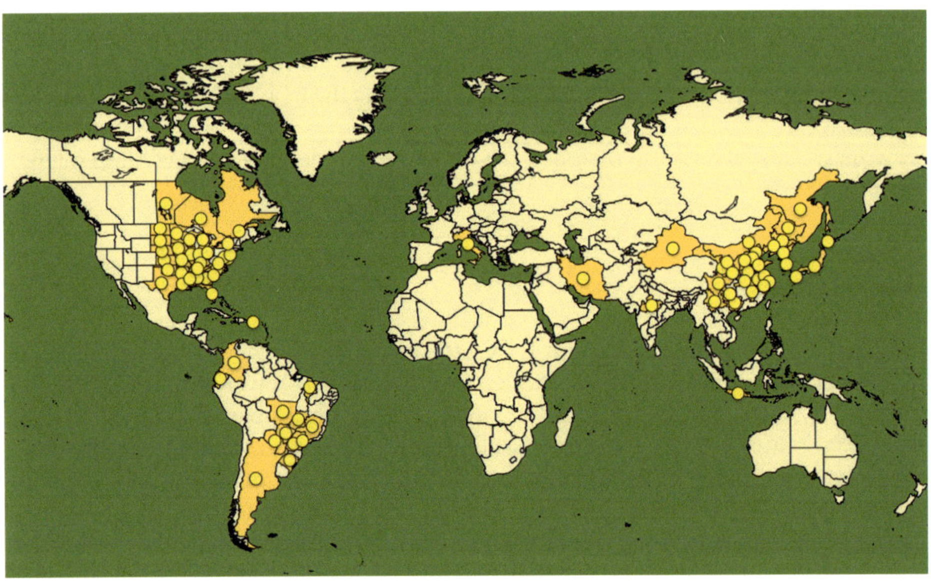

Figure 1.
Geographical distribution map Heterodera glycines. The yellow areas and circles correspond to regions of the world where the nematode is currently present. Source: EPPO global database (2023).

been introduced to the market over the last decades. This is the case of tioxazafen, Fluopyram, and fluazaindolizine, which are considered new-generation nematicides as they are less toxic to mammals and the environment, suggesting that they have mechanisms of action distinct from those observed for CMs and OPs types.

2. PPNs: overview

PPNs are microscopic and ubiquitous organisms that live in soil and plant tissues. Despite their inability to travel great distances, anthropogenic actions can lead them to be transmitted via contaminated plant material, soil, and agricultural machinery [11]. In addition to preventing plant roots from absorbing nutrients, PPNs may make plants more vulnerable to secondary phytopathogens such as fungi, bacteria, and viruses [16].

Many agriculturally important PPNs are members of the Tylenchida order, which includes endoparasites from the Heteroderidae and Pratylenchidae families. The Heteroderidae family includes the most important agricultural genera: RKNs (*Meloidogyne* spp.) and CNs (*Heterodera* and *Globodera* spp.). The family Pratylenchidae, on the other hand, comprises migratory endoparasite RLNs (*Pratylenchus* spp.), with over 60 species documented. All of the species in those genera are capable of infecting over 3000 distinct plant species, making them a global threat [17, 18].

PPNs, unlike free-living nematodes, have a hollow and protractile stylet that allows them to pass through plant cell walls, but not the plasma membrane, enabling them to feed on cell contents via the feeding tube that is solely connected with the stylet [19]. PPNs release a cocktail of polymeric or non-polymeric effector proteins that degrade and modify cytoplasmic components, causing the host cell's physiology and morphology to change [20]. The force for the nematode to feed on the cellular contents is provided by a muscular pump in the basal nodes (metacorpus). In this process, a complex plant-parasite interaction occurs, involving a sequence of recognition and response events that vary depending on the kind of parasite and host plant [21]. Although many aspects of the pathogenesis process are unknown, it is evident that acquiring this information is necessary for the creation of new and more effective techniques of treatment and control.

3. Biology and parasitism of the main PPNs of agricultural importance

3.1 Root-knot nematodes (*Meloidogyne* spp.)

RKN are obligate biotrophic endoparasites that require a host plant to complete their life cycle [20]. There are more than 90 species, but only a few have worldwide agricultural value. The most damaging species include: *Meloidogyne incognita*, *M. javanica*, *M. hapla*, *M. enterolobii* and *M. arenaria* [22]. They generally feed and reproduce inside the roots of plants, where they promote the establishment of permanent feeding sites, which subsequently give birth to galls or root-knots. Because these root-knots can be mistaken with those formed by other PPNs such as *Nacobbus aberrans* and *Subanguina radicicola*, their identification and diagnosis are essential, as Meloidogyne is generally more aggressive than the latter [22].

The RKNs life cycle starts when an adult female deposits their eggs (~500) in gelatinous-textured sacs (egg masses) generated by six anal glands (**Figure 2**).

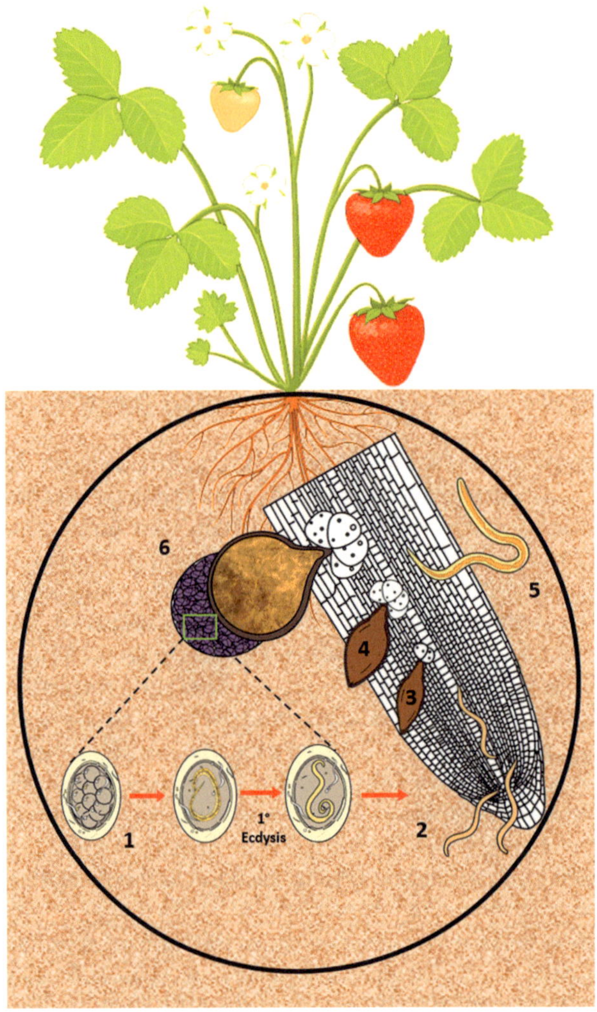

Figure 2.
Schematic representation of the root-knot nematode life cycle. 1: Egg in development, 2: J2 larvae entering root, 3 and 4: Swollen nematodes feeding in root, 5: Free male on the ground, 6: Mature female breaking through root surface with egg sac.

This gelatinous sac is made up of glycoproteins, which, in addition to protecting the egg, serve as a regulator of embryonic development by detecting changes in the temperature and humidity of the environment. With the advance in feeding, the female becomes more voluminous and the posterior section of her body is extended in such a way that allows the egg mass to be exposed outside the root, thus forming the nodes or galls [23]. The galls make it easier for microorganisms in the rhizosphere to come in contact with the eggs; however, their gelatinous structure has been shown to have antimicrobial properties that protect them from attack by microorganisms [24].

After embryogenesis, J1 undergoes the first molt (1° ecdysis), which will give birth to the infective form known as the second instar juvenile (J2). When temperature and humidity conditions are favorable, hatching of the eggs occurs, releasing the J2 nematodes. Unlike other PPNs, which require stimuli from plant roots for hatching, *Meloidogyne* spp. hatching can occur with or without plant stimuli [25]. In the soil, J2 carries out a pre-parasitic motile phase. During this period, it is susceptible to biotic and abiotic stresses, surviving, before infecting a root of a susceptible host, at the expense of its lipid reserves [26]. *Meloidogyne* spp. stages that can be discovered in the soil include J2 and mature males. Although the mechanisms by which J2 finds a susceptible plant are not fully understood, it is known that plant root exudates contain substances such as carbon dioxide, tannic acids, flavonoids, and volatile organic acids that may regulate J2 arrival or leave events [27, 28]. In addition, PPNs can be attracted by chemicals released during the attack by herbivores or insects. Thus, for example, it has been shown in *Tylenchulus* spp. that they are more easily attracted to plants infested by insect larvae than to non-infested plants [29].

J2 enters and migrates via the vascular cylinder of a vulnerable plant after penetrating very close to the root tip where there is a lack of completely developed endodermis (elongation zone) [30]. During this action, the intermediate lamella softens mechanically and enzymatically [31]. After traveling a short distance into the plant, J2 becomes sessile due to the atrophy of its somatic muscle (except for the head). The head is buried in the periphery of the vascular tissue cells (zone of cell differentiation), where it feeds (through a feeding tube) on the protoxylem and protophloem contents before the cells differentiate into specialized nurse cells known as giant cells (GCs) [32]. After the establishment of a permanent feeding site, the nematode becomes sedentary, growing significantly in size until it reaches maturity.

Although the mechanisms by which GCs are formed are not fully understood in detail, it is known that, during feeding, parasites secrete effector proteins that modify cell division events through reorganization of the cytoskeleton and signaling pathways [33]. This cocktail of effector molecules, which facilitate the establishment of the feeding site, is mainly produced by the dorsal gland. Peptides, virulence factors, β-1,4 endogluconases, polygalacturonases, xylanases, chorismate mutase, among others, are some of the molecules that compose the biochemical arsenal of the RKNs secretome [34].

The cell wall between the daughter nuclei is created in telophase during normal plant cell division; however, in the creation of GCs, the cell splits without cytokinesis and hence without cell wall development. This causes multinuclear GCs to form, as well as the proliferation of other organelles such as endoplasmic reticulum, mitochondria, and ribosomes [35]. These metabolic and physiological changes give GCs the characteristics of metabolically active cells, with the inner growth of the wall forming an extensive labyrinth network that increases the surface area of exchange with the associated membrane of xylem vessels [36]. Although less visible, phloem sieve components develop around large cells to ensure phloem continuity [37].

Compared with uninfected plants, tissue from RKNs-infected plants shows a significant increase in the phytohormone auxin [38]. Auxins, like ethylene, have a role in the formation of cell wall ingrowths by promoting the production of cell wall modifying proteins as well as proteins implicated in acidic growth (proton pumps) [39, 40]. These and others evidence imply that the parasite's modulation of this phytohormone is a key factor in the creation of GCs. In addition to auxins, cytokinins, which are involved in cell division, are elevated in infected tissue, suggesting that they may actively contribute to the parasite-feeding site formation [41].

The plant defense system is also affected by the effector proteins controlling the phytohormone pathway: salicylic acid (SA) and jasmonic acid (JA). In general, the SA pathway protects against biotrophic and hemibiotrophic pathogens such as endoparasitic sedentary [42]. The JA pathway, on the other hand, is involved in plant defenses against necrotrophic pathogens and leaf-chewing insects [43].

The capacity of biotrophic and hemibiotrophic pathogens to modify plant defenses is important to host tissue colonization success. The first line of activated protection against RKNs is the pathogen-associated molecular pattern (PAMPs) triggered immunity (PTI) [44]. In response to the attack (PTI response), the plant triggers the synthesis of reactive oxygen species (ROS), as well as the production of protein kinases (MAPKs) and the activation of the JA and SA pathways [45]. SA regulates the expression of several genes involved in the synthesis of proteins mediating the pathogenic response, while the JA pathway regulates the expression of genes encoding for proteins involved in the synthesis of thionin, defensin, and phytoalexin, among others [45, 46]. The level of response and plant-parasite interaction seems to be influenced by the type of plant, as well as by the type of parasite and its infective state [20].

After the establishment of the feeding site, J2 undergoes significant morphological changes, including two successive molts (J3 and J4). Unlike J2, these last two stages do not feed, as they lack a functional stylet [47]. Under appropriate environmental conditions and with sufficient food availability, J4 undergoes a final ecdysis, the female acquires the classical pear shape (500 times the volume of J2) and the male is left free in the environment. Before reaching full maturity, the cells of the genital primordium of the female divide to give rise to the rectal glands, which will later secrete the gelatinous material with which she will protect her eggs. On the other hand, the male characteristics (testis and vas deferens) differentiate from the posterior end, where they connect with the rectum [25].

It has been recorded that, within the genus Meloidogyne, there are three different forms of reproduction. On the one side, there is the fertilization of oocytes by the male (amphimixis), while, on the other hand, there is parthenogenesis, which can be meiotic or mitotic. In the first case, there is a fusion of the pronucleus of an ovule, generated by meiosis, with a polar body (automixis); while, in the second case, the oocyte divides by mitosis, where one of the nuclei deteriorates and the other remains to give rise to the embryo [48, 49].

As in CNs, sex chromosomes are absent, so sex is strongly influenced, at least in parthenogenetic species, by environmental conditions as well as by host plant conditions. Lack of food, overcrowding, extreme temperatures, and soil dryness are some of the conditions that favor the development of males that reproduce by meiotic parthenogenesis [50]. These males rarely fertilize females, and when they do, mitotic parthenogenesis takes place without any fusion with the sperm nucleus.

3.2 Cyst nematodes (*Heterodera* and *Globodera* spp.)

After RKNs, the CNs constitute the second group of sedentary endoparasites with the greatest impact on world agriculture. Among the most important species are *Heterodera glycines*, also known as the soybean cyst nematode (SNC); *Globodera pallida* and *G. rostochiensis*, both known as potato cyst nematodes (PCNs); and, finally, nematodes that attack cereals such as *Heterodera avenae* and *H. filipjevi* [22].

Like RKNs, the life cycle comprises an egg stage, four juveniles (J1,2,3,4), and adults (male and female). However, there are important differences that will be discussed below.

Unlike Meloidogyne, in CNs, the eggs may be retained in the female's body, or, depending on the species, deposited in a gelatinous sac. After their deaths, the female body transforms into the cyst, which protects and holds the eggs inside. Depending on its maturation stages, it can acquire different shades ranging from black-brown to yellow. These color changes are known to be due to the activity of the enzyme polyphenol oxidases, which, in addition to catalyzing the hydroxylation of phenolic compounds, also catalyze the polymerization of o-quinones on the cyst cuticle [51].

In the cyst, the embryo can remain for many years in a state of metabolic suspension (dormancy stage) when environmental conditions are unfavorable. This dormant state is essential for the survival and pathogenesis of the CNs. In general, normal embryo development requires the action of internal signaling molecules, which are regulated by biotic and abiotic conditions [52]. Not all species have this dormancy stage in their life cycle; however, for those that do, it can be of two types depending on the dormancy strategy used: quiescence and diapause. Quiescence can be an obligate or facultative response to poor environmental conditions, which is reversed when external conditions are ideal. On the other hand, diapause, which can also be obligate or facultative, is a state in which development of the embryo (J1) is completely arrested until the metabolic requirements are not satisfied, so that good environmental conditions are not sufficient to reverse the process [53].

Although the egg may hatch spontaneously, there are, however, factors that stimulate hatching. These may be environmental (temperature, soil moisture, oxygen availability, etc.) or derived from root exudates [54]. The effects of such factors on egg hatching vary between species of CNs [55]. The hatching of *H. glycines* and *G. rostochiensis* eggs is partly dependent on factors derived from root exudates. Terpenoids (glycinoeclepin A and solanoeclepin A, B), glycoalkaloids (α-solanine and α-chaconine), metavanadate, picrolonic acid, and phenanthroline derivates, are some of the factors identified in root exudates that stimulate egg hatching [56–58]. These factors can be used as agrochemicals in the control of PPNs, since they can be applied in infested fields, stimulating the release of J2 before planting the crop.

Although the exit of J2 from the cyst constitutes the last stage of the hatching process, it is known that for this to occur, changes in the permeability of the membranes that cover the egg must take place. These changes are essential since they trigger the activation of second messengers that promote the following stages of development, such as the activation of metabolic pathways and J1 ecdysis [59]. Three membranes are known to surround the egg. The inner one is formed mainly by lipids and is semi-permeable to water and small ions; the intermediate, constituted by chitin microfilaments that give resistance and flexibility to the egg; and, finally, an outer one, is formed by lipoproteins (vitelline layer), which are believed to be essential for egg fertilization [60].

Unlike RKNs, where J2 activation can occur before changes in eggshell structure, in CNs, changes in eggshell membrane permeability are required for J2 activation. Within the egg, the movement of the juveniles is reduced by the low turgor pressure of the perivitelline fluid. It is mainly composed of the disaccharide trehalose (α-D-glucopyranosyl 1–1, α-D-glucopyranoside), which not only serves as an energy source for the nematode but also exerts a high osmotic pressure that reduces the water content of the perivitelline fluid [61]. Reduction of the trehalose content, and thus hydration (increased turgor pressure), is a prerequisite for egg hatching [60]. Although this is true for many species, there are others (i.e., *Heterodera schachtii*) that, on the contrary, have a low osmotic pressure of the perivitelline fluid allowing them to hatch under osmotic stress conditions [62]. Changes in the permeability of eggshells to Ca^{2+} ions, modifications in their lipoproteins, as well as the presence of zinc-dependent enzymes that mediate hatching, are other factors known to be involved in such phenomena [61, 63]. After the first ecdysis, J2 becomes metabolically active and uses its stylet to exit the egg. Once out of the egg, it is believed that it leaves the cyst through the hole generated by the detachment of the head from the female's body once it has reached full cyst maturation [64].

Free in the soil, J2 is vulnerable to environmental conditions, so it must find a suitable root host plant before its lipid reserves are depleted. Under optimal conditions, localization time for *G. pallida* and *G. rostochiensis* has been estimated between 6 and 11 days [65]. In soil, J2 must orient in a three-dimensional matrix, responding to changes in the gradient of a wide variety of physical and chemical stimuli (i.e., CO_2, temperature, pH, redox potential, ethylene, phytohormones, etc.). Some are primarily involved in orientation, while others are in root attraction. It is generally accepted that these signals are perceived through amphids [66].

Once J2 reaches the root of a susceptible plant, it enters through the elongation areas of the root tip, as described for *Meloidogyne* spp. As in RKNs, penetration enzymes such as cellulases β-1,4 endogluconases are produced; however, there is a difference in their entry into plant tissues, as in CNs the movement is intracellular, while in RKNs it is intercellularly [67]. In CNs, the permanent feeding site is known as a syncytium (composed of about 200 cells). As in RKNs, its formation is a complex process that requires the presence of effector proteins modulating the plant-parasite response. At first, the parasite introduces its stylet into the initial syncytial cell (ISC) in order to evaluate the type of response. When the response is unsatisfactory (covering the stylet by callose), the nematode withdraws its stylet to continue trying with another cell close to the one initially selected. Once the ISC is selected, the maturation process of the syncytium begins, which will become its feeding site for life [68].

In general, the process of syncytium formation is a highly complex process, which, as in RKNs, involves changes in gene expression in both the plant and the nematode. After 7 hours of ISC preparation, the sub ventral glands atrophy while the dorsal gland becomes more active [69]. The stylet is then removed from the ISC and reintroduced. It is at this point that the effector proteins contained in the dorsal gland are released into the cytoplasm of the ISCs [68]. As mentioned above, the secreted proteins have the function of establishing syncytium, which is facilitated by the dissolution of the cell wall separating adjacent cells from the ISC [70]. In *Arabidopsis thaliana*, it has been observed that J2 of *H. schachtii* can select procambial or cambial cells for ISC formation. In the first case, the syncytium extends to the xylem and phloem vessels, whereas in the latter case, it maintains contact with the phloem through companion cells [71, 72].

Syncytial cells, like GC in RKNs, undergo morphological changes. The nucleus enlarges by endoreduplication, and the central vacuole breaks down and gives rise to a large number of small vacuoles; on the other hand, the cytoplasm expands to accommodate a large number of ribosomes, mitochondria, endoplasmic reticulum, etc. [68]. In addition to these morphological changes, cells also experience metabolic changes that involve the overexpression of genes that are part of primary metabolism. These changes are influenced by effector proteins that modulate the transport and function of phytohormones. Thus, for example, chorismate mutase, an enzyme isolated from several PPNs and phytopathogenic fungi, is implicated in the initial pathogenesis of infection since it is known to alter the shikimic acid pathway, which is essential for plant-nematode interaction, since through it, precursors for auxins, salicylic acid, and a wide variety of phenolic compounds are generated [72, 73].

After the establishment of the feeding site, the nematode develops to the adult stage. The determination of males and females is influenced by environmental conditions and the supply of nutrients from the host plant. Females, unlike males, have a higher demand for nutrients, so their syncytia are larger than those formed by males. As adults, males, unlike females, are mobile and leave the root to find and fertilize a female. On the other hand, the female enlarges considerably to maintain the eggs

(~600 inside the cyst and ~ 200 outside it) and thus restart the parasite's life cycle, which has been estimated at 4 weeks for *H. glycines* under favorable conditions [60].

3.3 Root-lesion nematodes (*Pratylenchus* spp.)

The RLNs are obligate biotrophic migratory endoparasites nematodes, without a permanent feeding site. There are more than 90 species registered with a cosmopolitan distribution. After RKNs and CNs, it is the third most important genus in agriculture. The species with the greatest agricultural economic impact are *Pratylenchus penetrans*; *P. thornei*; *P. neglectus*; *P. zeae*; *P. vulnus* and *P. coffeae* [22].

As migratory endoparasites, they move intra- and inter-cellularly through the root tissues, generating superficial cracking of the root with internal rotting of the tubers that predispose to secondary infections by fungi and bacteria. Unlike the infections generated by RKNs and CNs, where the galls and cysts are observed with the naked eye on infested plants, in RLNs it is not easy to identify signs or symptoms that warn about their presence [74].

Its life cycle is relatively simple. This can last between 3 and 6 weeks, depending on the environmental conditions and the host. The eggs are deposited by the female either in the soil, the roots, or the tubers. Like RKNs and CNs, J2 hatches from the egg, which later develops through J3, J4, and the adult stages: male and female. All are vermiform (worm-like), infective, and motile. This ability to remain mobile during all phases of their cycle allows them to freely enter and exit the host plant [75]. Males commonly reproduce by parthenogenesis, and in some species, they are frequent, while in others they are not [76].

Like other PPNs, they are attracted to root elongation zones by different chemotactic molecules released by the host plant. They have a short, robust, hollow-mouthed stylet, which they use to break the plant cell wall by repeatedly pushing it. J2 and J3 tend to feed on the root hairs; however, they can also do from the epidermal, cortical, or stellar cells [77]. Prior to ingestion of the plant cell's cytoplasmic contents, the dorsal glands are activated to secrete effector proteins that accumulate in the ampulla behind the buccal stylet. Feeding time can be short or last a few hours. In the latter case the cell may die, however, during the brief feeding, the stylet is removed, and the puncture site is sealed with no contents leaking out. Also, their intracellular displacement is another cause of cell death even when feeding occurs for short periods of time [78, 79].

Initial invasion is followed by the penetration of other nematodes, attracted by compounds released from damaged cells. Migration together with the feeding points generates brown lesions (necrotic areas,), which vary according to the type of host plant and parasite species. This roam feeding behavior leads to loss of plant growth and the appearance of leaf chlorosis and swelling and atrophy of the roots [80].

Although not all aspects of the biology of parasitism in RLNs are known, it is known that many of the effector proteins are different from those produced by RKNs and CNs. It is to be expected that, due to their migratory behavior and their intermittent feeding, the arsenal of effector proteins to manipulate the response of the host plant is less extensive than that of the sedentary endoparasites. Thus, for example, no orthologous sequences have been found for effector proteins essential for the establishment of infection in sedentary endoparasites such as 19C07 (auxin influx in syncytium], chorismate mutase, 10A06 (antioxidant genes in syncytium), CLE peptide, auxin, peptide hormones (CEPs), cytokinin, among others [74]. *In silico* studies of alignment between conserved domains of effector protein sequences have shown that these are different among sedentary nematode species. Thus, for example, it has been shown that among RLNs and RKNs, there are only a few effector proteins in common, among which cysteine proteases, pepsin inhibitor-3, astacin, and a domain of unknown function called DUF148 stand out. Although the function of this domain is not clear, it is believed that it may be part of the SXP/RAL-2 (ANIS5) family of proteins that bind cations such as calcium and magnesium [81].

Transcriptomic studies in *P. thornei* and *P. coffeae* have provided evidence of the secretion of antioxidant enzymes such as superoxide dismutase, glutathione peroxidase, and peroxiredoxin, which block the action of reactive oxygen species (ROS) produced by the host plant. On the other hand, the resistance of some cultivars to infection is due to the presence and accumulation of secondary metabolites; however, due to their ability to move freely through plant tissues, many of the RLNs escape the defense activities of the host plant [74, 82].

Like other nematode species, *Pratylenchus* spp. have acquired the ability to remain in the anhydrobiotic state for long periods of time; however, their success rate is moderate compared to other nematodes such as *Dytilenchus dipsaci* or *G. rostochiensis* [83]. The rate of loss of soil moisture (dehydration), the stage of development of the parasite, the presence of plant residues (root and stem), as well as genetic characteristics, are some of the factors that affect the development of anhydrobiosis of *P. penetrans* in dry soil [84]. Unlike J4, which is resilient, the other forms of the parasite are sensitive to changes in soil moisture. Eggs, on the other hand, survive changes better than motile forms of the parasite [85]. In addition to withstanding dry soil, these can also survive freezing as anhydrobionts. The loss of water from their bodies reduces the formation of ice needles that can pierce internal organ cells [86]. Although the reproductive capacity is not affected by anhydrobiosis, the reserve lipid content is managed differently during this stage, since it has been shown that lipid consumption is higher in females than in males [85, 87].

4. Management: a focus on new nematicides

4.1 Overview

The management and control of PPNs are of crucial importance in agriculture since many of them can endanger the food security of a country due to the loss of productive yield of the crops. After the USA and China, Brazil is the third country with the highest grain production in the world. According to the FAO [88], Brazil leads world soybean production, with 134 million tons produced in 2021, 10% more than that produced by the USA: ~120 million tons. It is estimated that 20% of soybean production in Brazil is affected by *H. glycines*, which constitutes a phytosanitary problem that generates economic losses of more than R$16 billion per year [89].

Although crop rotation, genetic improvement, and biological control are some of the alternatives currently available for the management and control of PPNs (**Figure 3**), chemical control is the most widely used method worldwide [90, 91]. In developing countries, where access to agricultural technology is limited, and where the cost-benefit factor is of paramount importance for the economy, the use of agrochemicals is still the method of choice, due to their ease of implementation and availability [92, 93].

Almost half of the global nematicides market, worth US$ 1.3 billion, is used to control species of the genera *Meloidogyne* and *Heterodera* [94, 95]. However, with

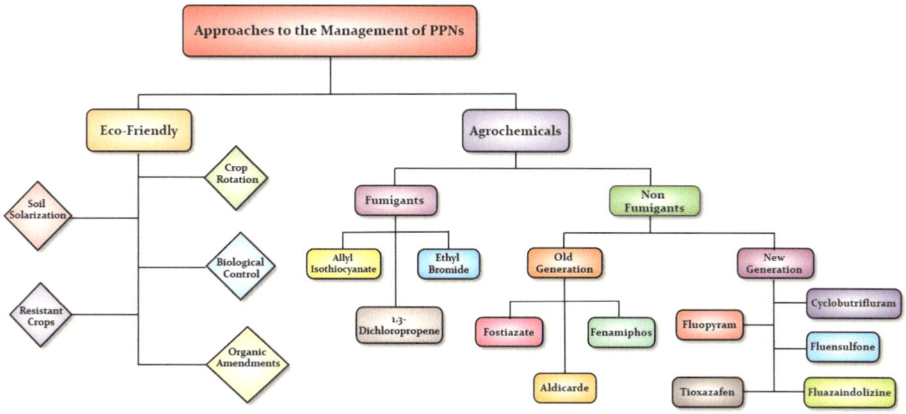

Figure 3.
Flow chart summarizing the different approaches that currently exist for the management and control of PPNs.

increasing requirements for food safety and environmental protection, several nematicides are no longer considered suitable for modern agriculture, as there are much data showing their high toxicity to the environment as a whole. For example, the use of methyl bromide and 1,3 dichloropropene (fumigant nematicides) are currently banned for the control of PPNs [96, 97]. On the other hand, non-fumigant nematicides such as organophosphorus (OPs) and carbamates (CMs), are also prohibited in many parts of the world, due to their high toxicity to humans and the environment [98].

OPs and CMs owe their high toxicity to their low selectivity in their mechanism of action, since they are reversible inhibitors of the cholinesterase enzyme (AChE). This is because many are used as broad-spectrum pesticides (insecticides, acaricides, nematicides, and rodenticides) and as chemical weapons [99]. Aldicard, carbofuran, fenamiphos, and oxamyl are some of the OPs and CMs in which cholinesterase inhibition of PPNs has been demonstrated. Since the concentration of nematicide in agricultural soils is not high enough to kill nematodes, many of these compounds exert a nematostatic function with temporary paralysis that interferes with host feeding and egg hatching [100].

Although resistance has been reported under controlled laboratory conditions in PPNs, there are no proven examples in the scientific literature documenting dramatic changes in tolerance to nematicides that lead to suspicions about the development of resistance under field application conditions [100]. On the other hand, field resistance to agrochemicals has been widely demonstrated in different species of agricultural pests; however, it is believed that in PPNs the risk of developing resistance under field conditions is theoretically unlikely [101]. The arguments supporting this hypothesis are based on the application methods of nematicides and the ecological behaviors of PPNs. These characteristics reduce the effects of selective pressure to develop resistance, which can be overcome by the indiscriminate use of nematicides with different mechanisms of action [102].

Despite the fact that the global market for nematicides is much smaller than that of other agrochemicals such as insecticides (US$16.4 billion in 2019), fungicides (US$13.4 billion in 2019), and herbicides (US$32.6 billion in 2019), the fact remains that the market continues to grow despite strong regulatory rules limiting their use. Thus, for example, in 2011 dividends were approximately US$1 billion; however, by

2022 it was US$1.78 billion, an increase of approximately 43% [95, 100]. Post-modern control of PPNs requires cost-effective and selective targeted control strategies that meet the environmental safety demands of both growers and consumers. Thus, in the last 10 years, new nematicides have emerged (**Figure 4**) with low persistence in the environment and with selective mechanisms of action different from those of OPs and CMs. Of these new nematicides, only fluopyram and cyclobutrifluram have a currently known mechanism of action on PPNs [103, 104].

4.2 Fluopyram and cyclobutrifluram

Marketed as an active ingredient in Verango®, Velum® and Indemnify®, fluopyram (N-[2-[3-chloro-5-(trifluoromethyl)pyridin-2-yl]ethyl]-2-(trifluoromethyl) benzamide) is a new generation nematicide-fungicide developed by Bayer Crop Science in 2012. Like fluopyram, cyclobutrifluram is a trifluoromethylpyridine developed and pantented by Syngenta in 2013 and that is marketed under the names Vaniva®, Tymirium®, and Victrato® [105]. Both interfere with the electron transport chain of the mitochondrial system (complex II) through inhibition of the enzyme succinate dehydrogenase (SDH). This mechanism has been reported for both PPNs and plant pathogenic fungi [106].

In eukaryotic organisms, SDH catalyzes the oxidation of succinate to fumarate as part of a mechanism to transfer electrons through the quinone pool. On the other hand, in nematodes and anaerobic helminths, complex II is associated with another enzyme that performs the reverse reaction (oxidation of reduced quinones) known as fumarate reductase (FRD). This is an adaptation to low oxygen pressures that prevent completely aerobic respiration. Such conditions are typical in intestinal nematodes; however, in free-living nematodes such as *Caenorhabditis elegans*, FDR is present as well as SDH, suggesting that FDR is an inducible enzyme under anaerobic conditions [107].

SDH consists of four subunits (A-D) that differ in structure and function. Thus, for example, the SDHA sub-unit is the largest and is where the active site of the enzyme is located. On the other hand, the SDHB subunit contains iron-sulfur groups that mediate electron transfer to ubiquinone, which is located at the interface between the SDHB, SDHC, and SDHD subunits. It is there that many SDHIs bind to interfere with the catalytic activity of complex II [107]. Sequence alignments between

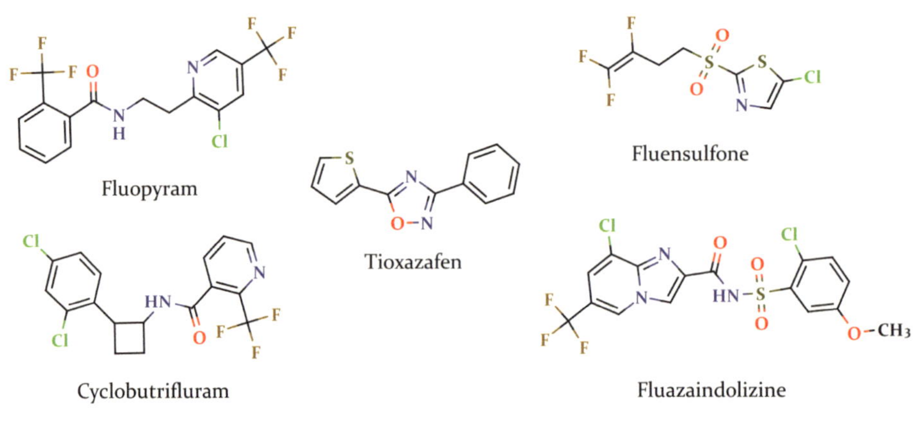

Figure 4.
Chemical structures of new nematicides.

sub-units of *C. elegans* and *Meloidogyne* spp. show that there is a high degree of similarity [108, 109]. It has been shown in *in vitro* studies that fluopyram binds to complex II of *C. elegans* with an inhibition constant (Ki) of 2.0 nM [110]. Additionally, it has been possible to generate *in vitro*, mutants resistant to fluopyram and other SDHI [100]. Although more expensive than other SDHIs such as flutolanil and solatenol (US$650/ha approx.), fluopyram is the only nematicide that is effective for both *C. elegans* and PPNs [103, 111]. Studies on mammalian SDHs show that fluopyram has little effect on the activity of these enzymes, suggesting that it is a compound that is highly selective for nematode SDHs. These findings explain the toxicological profile of fluopyram, which is considered, at moderate doses, safe for other life forms [110].

Compared to fluensulfone, fluopyram works fast-acting. At a concentration of 2.0 mg/L, 100% of the J2 of *Meloidogyne incognita* are immobilized after 24 h of exposure *in-vitro*. At the same concentration, 48 hours are required for fluensulfone to immobilize 100% of the nematodes [112]. The half maximal effective concentration (EC_{50}) required to immobilize J2s of *M. incognita* is 0.7 mg/L, which is comparatively lower than that achieved by fluensulfone (50 mg/L) [100, 94]. Despite its strong nematicidal activity against *M. incognita*, fluopyram is nematostatic against *Heterodera schachtii* at a concentration of 20 mg/L [113]. It has also shown good in-vitro activity against *R. reniformis* and *Ditylenchus dipsaci* with visible effects after 2 hours of exposure at a concentration of 5.0 mg/L and 3.0 mg/L, respectively [103, 114].

Unlike fluensulfone, abmectin, and aldicard, fluopyram has a poor ovicidal effect. At a concentration of 2.5 mg/L, they have a slight effect on the inhibition of the hatching of *M. incognita* eggs [104]. On the other hand, an 81% reduction in the inhibition of *G. pallida* egg hatching has been observed in vitro at a concentration of 5 mg/L. However, they are easily recovered after washing. At concentrations greater than 50 mg/L hatching inhibition is achieved without recovery [115]. In greenhouse and field trials on tomatoes, carrots, cotton (*Gossypium hirsutum*), potato, and soybean, fluopyram has been effective at managing PPNs [116–118].

4.3 Fluensulfone

Fluensulfone (5-chloro-2-(3,4,4-trifluorobut-3-enylsulfonyl)-1,3-thiazole) is the active ingredient of Nemitz®, a nematicide belonging to the group of fluroalkenyl sulfones, used for the control of PPNs in a wide variety of agricultural crops [119]. First registered in the USA in 2014 [120], it was developed by ADAMA and Control Solutions for Quali-Pro brand, in the past Makhteshim Agan Industries Ltd. It exists in three commercial formulations, one for turfgrass (Nimitz® Pro G) and two for horticulture: granules (Nimitz® 2% GR) and emulsion (Nimitz® and Nimitz® 480 EC).

Unlike OPs and CMs, fluensulfone is less toxic to humans and non-target organisms. It is moderately toxic after oral exposure, and has low acute toxicity following dermal application or inhalation in rabbits and rats [121]. The lethal dose via oral administration (LD_{50}) in rats is 671 mg/kg body weight, which is five hundred times safer than the old nematicides such as aldicard (1 mg/kg of body weight) [122]. Despite the above, it has been reported to be toxic to aquatic organisms. Thus, for example, the EC_{50} for *Daphnia magna* is 0.35 mg/L after 48 h of exposure and 0.04 mg/L for *Pseudokirchneriella subcapitata* after 72 h of exposure [123]. In addition, fluensulfone is phytotoxic, thus it should be applied 7 days before planting [124]. Unlike soil fumigants, there is no need for a limiting buffer zone, as the re-entry time is essentially only 12 hours after application.

Although there are few studies that demonstrate the impact of fluensulfone on free-living nematodes, it is known to be less toxic to *C. elegans* than to *M. incognita* [125]. Unlike fostiozate, fluensulfone has been reported not to affect the biodiversity of free-living nematodes [126]. In tests on turfgrass soil, fluensulfone has been observed to have a lower impact on the density and biodiversity of free-living nematodes when compared to fluopyram and abamectin; however, it reduced the green grass cover [127].

The immobilization in the form of a straight rod generated by fluensulfone when applied to J2 differs from that produced by OPs, suggesting a different mechanism of action than AChE inhibitors [128]. However, the exact mechanism of action on PPNs remains unknown. Some hypotheses suggest that fluensulfone possibly inhibits the enzyme acyl-CoA dehydrogenase, which is key in mobilizing lipids to obtain energy [129]; however, others suggest that its mechanism of action may be related to that of fluopyram [100, 103]. In insects, the difluoroalkenyl derivatives were found to inhibit β-oxidation of fatty acids in the mitochondria [130].

Compared to OPs and fluopyran, fluensulfone produces a slower but irreversible paralysis. *In vitro* tests show that after removal with water of fluensulfone from J2 *Meloidogyne hapla* exposed to 1.0 mg/L for 24 h, 90% remained immobile [131]. In addition, J2 of *M. incognita* exposed for 17 h to 4.0 mg/L fluensulfone lost the ability to infect lettuce seedlings after rinsing with water [112]. This loss of infective capacity has also been reflected in the number of galls that form when plants are infected with nematodes that have had fluensulfone removed with water.

Like other nematicides, the susceptibility of PPNs to fluensulfone depends on the genus and species, even registering variations between the same species. In migrating nematodes such as *Ditylenchus dipsaci*, *Bursaphelenchus xylophilus*, and *Aphelenchoides* spp., it has been shown that they are more tolerant to fluensulfone than *M. javanica* [132]. While J2 of *M. javinica* is immobilized at a concentration of 0.25 mg/L for 48 h, in *B. xylophilus* and *D. dipsaci* a concentration higher than 16 mg/L is required for complete immobilization of these nematodes after 48 h of exposure [131, 132]. Among nematodes of the genus Meloidogyne, it has been seen that *M. javanica* is more tolerant than *M. incognita* and *M. hapla*. Thus, for example, the EC_{50} for *M. javanica* after 48 h of exposure is 0.83 mg/L; while for *M. incognita* and *M. hapla* it is 0.12 and 0.41 mg/L, respectively [100, 128]. Although its ovicidal effect is better than that of fluopyram, it is limited, requiring concentrations greater than 50 mg/L for a 50% reduction in hatching of *M. incognita* eggs; however, its effect is irreversible [115].

Currently, fluensulfone is approved for the control of different PPNs species (Meloidogyne, Pratylenchus, Hoplolaimus, Globodera, etc.) in a wide variety of crops such as tomato, potato, pepper, cabbage, squash, and strawberry, among others [100, 133]. Depending on the number of nematodes (population density per gram of soil), the type of crop rotation, and the presence of resistant plants, the average application rate is between 1.9 and 3.3 kg/ha [100, 129]. There are different forms of application; however, the most recommended are drip injection or broadcast by mechanical incorporation with a single application 7 days before planting [129]. It has been estimated that its half-life in soil is 36 days, with a dissipation rate (DT_{50}) of between 20 and 50 days, which can vary depending on the type of soil and environmental conditions [134, 135].

4.4 Tioxazafen: seed nematicide

Registered in 2017 as a seed nematicide (NemaStrike™ ST), tioxazafen is a phenyloxadiazole (formally 3-phenyl-5-thiophen-2-yl-1,2,4-oxadiazole) developed by Monsanto Company, now Bayer CropScience, for the management and control of nematodes in cotton, corn, and soybean crops [136]. Unlike other nematicides, the development of tioxazafen involved the use of *in silico* approaches such as computational screening, scaffold hopping, and ligand-based model. After evaluating the nematicidal activity of different models of a series of stilbene, chalcone, and azobenzene compounds (SCA series), tioxazafen emerges as the best candidate for combining intrinsic efficacy, longevity in soil, and synthetic accessibility [137].

Although its mechanism of action is not known with certainty, it is hypothesized that its nematicidal activity is due to ribosomal disruption. Polymorphism studies on *C. elegans* mutants that are tolerant to high doses of tioxazafen have revealed that resistance is caused by a single nucleotide change in the gene coding for the L3 subunit of the mitochondrial ribosome [138]. On the other hand, *in silico* studies of the prediction of the target of small molecules through virtual screening, modeling, and molecular docking indicate that another possible mechanism of action of tioxazafen is through the inhibition of chaperone proteins [Arley, unpublished].

It is known through *in vitro* bioassays and X-ray crystallographic studies that benzisoxazoles and oxadiazoles are potent selective inhibitors of Hsp90 chaperone proteins [139, 140]. The Hsp90 protein is an essential chaperone in the folding of several proteins known as clients such as protein kinases, transcription factors, and hormone receptors, among others [141]. In nematodes such as *Brugia pahangi*, *Brugia malayi*, *Schistosoma mansoni,* and *Ancylostoma caninum*, Hsp90 is essential for survival inside the host [142, 143]. In *C. elegans*, Hsp90 is essential for larval development (L2-L4) and nematode longevity, since its chaperone activity stabilizes the nuclear transcription factor DAF-16/FOX, responsible for activating genes involved in longevity, lipogenesis, oxidative stress, glycolysis, innate immunity, and reproduction [144].

Incorporated as an active ingredient in fungicide-insecticide formulations (Acceleron®), tioxazafen has low water solubility (0.0125 mg/mL) and reduced mobility in soil with an octanol/water partition coefficient (Log P) of 3.26. These characteristics make tioxazafen a moderately lipophilic compound that, according to estimates, has an ambient half-life of more than 100 days under aerobic conditions [138]. According to *in vitro* and *in vivo* studies, tioxazafen and its derivatives (thiophenic acid, benzamidine, etc.) do not pose any risk to terrestrial invertebrates, amphibians, reptiles, pollinating insects, and aquatic plants. However, it has been shown that consumption of tioxazafen-treated seeds can be harmful to birds and some terrestrial vertebrates [145]. In tests using radiolabeled tioxazafen on soybeans, the chemical was shown to be dispersed mostly in the root area without penetrating the vascular tissues [137].

In studies in growth chambers and in the field, it has been possible to evaluate the larvicidal and ovicidal power of tioxazafen, as well as its ability to inhibit reproduction and colonization of the host plant. The results suggest that tioxazafen exerts a direct nematicidal action on PPNs at doses below those required for OPs and CMs (50 mg/L) [136, 138]. The EC_{50} for M. incognita and *R. reniformes* has been estimated to be approximately 60 mg/L after 24 hours of exposure. No recovery of motility was observed after the removal of tioxazafen with water. In bioassays, its ovicidal power is reached at concentrations above 3.0 mg/L [136].

Trials in growth chambers and greenhouses show that tioxazafen is an excellent controller of several key nematodes when applied to seeds of soybean, corn, and cotton. In field trials with corn and soybean cultivars conducted by the USA EUP (Experimental Use Permit) in 40 locations that had nematode populations three times greater than those needed to cause economically significant damage, it was observed that tioxazafen increased the productive yield more efficiently than those observed for commercial use nematicides [137].

In Brazil, the nematicidal power of tioxazafen on *Heterodera glycines*, *M. incognita*, *M. javanica*, *P. brachyurus*, and *P. zae* has been reported in soybean, corn, and cotton crops [146]. According to the results, to achieve a control equal to or greater than that of the positive control (imidacloprid + thiodicarb), 0.250 mg per seed is required to control *M. javanica* and *P. brachyurus*; while for the control of *H. glycines* and *M. incognita*, 0.500 and 0.750 mg per seed is required, respectively. When compared to untreated plants, phytotoxicity experiments on house vegetation demonstrate that tioxazafen has no effect on plant growth or biomass [146]. In tomato plants, however, it has been discovered that its larvicidal effectiveness is dependent on the concentration and species of PPNs, with *Pratylenchus* and *Rotylenchus* spp. requiring a larger dose of the product [146]. Although the efficiency of tioxazafen has been widely demonstrated both *in vitro* and in field studies, since 2020 tioxazafen has been voluntarily withdrawn by the manufacturer; however, tioxazafen is easily accessible synthetically, which is why it continues to be used in research on the rational design of bioactive compounds [136, 147].

4.5 Fluazaindolizine: selective nematicide

Corteva Agriscience™, an agricultural division of the DowDuPont company (formerly DuPont Crop Protection), discovered and developed Salibro™, based on the new active substance fluazaindolizine (Reklemel™ Active). It is the first member of the new chemical class of N-phenylsulfonylimidazopyridine-2-carboxamide nematicides [148]. Fluazaindolizine, which was first registered with the EPA in July 2021, is a pesticide used to control PPNs on crops such as carrots, squash, tomatoes, eggplant, potatoes, and taro, as well as some fruits such as oranges, peaches, almonds, and grapes [149]. Their discovery began with high-throughput screening of an internal compound library against RKNs. After evaluating different 2,5-disubstituted arylsulfonamides, fluazaindolizine was selected due to its ecotoxicological and nematicidal profile [148]. It is suitable for a broad range of application techniques, including drip irrigation, bed spraying, and soil incorporation, among others [150].

In vitro studies on targets associated with old nematicides such as AChE, mitochondrial electron transporters, glutamate-regulated chloride channels (GluCl), and nicotinic receptors (nAChRs); show that fluazaindolizine has a mechanism of action different from that of other nematicides [148]. However, its exact mechanism on PPNs remains unknown. After five days of exposure at a concentration of 30 μM, no significant impact was observed on the motility and mortality of non-target organisms such as *C. elegands*, *Drosophila melanogaster*, and *Diabrotica undecimpunctata*. On the other hand, *in silico* studies of structure-activity relationship show that a possible mechanism of action of fluazaindolizine is through the inhibition of the enzyme pantoate synthase [Arley, unpublished]. The enzyme is responsible for the synthesis of pantothenate (vitamin B_5), which is the precursor of coenzyme A (CoA), which in turn is an essential cofactor for the synthesis and degradation of fatty acids, as well as for the TCA cycle [151]. In PPNs there is a strict requirement for this vitamin, thus, for example, it has been observed that there is a horizontal transfer of genes for the synthesis of B complex vitamins (B_1, B_5, B_6, and B_7) between *H. glycines* and bacteria and fungi [152, 153].

Under *in vitro* conditions, it has been observed that the sublethal concentration of fluazaindolizine is between 5 and 250 mg/L. At concentrations greater than 500 mg/L, an acute poisoning of *Meloidogyne* is observed, without recovery after treatment with water [100, 148]. This loss of motility and infective capacity after treatment with water is affected even at concentrations of 5 mg/L [154]. The LC_{50} for RKNs is 177 mg/L at 72 h. The nematicidal effect of fluazaindolizine is both concentration and time-dependent. Additionally, it has been observed that there is a different susceptibility to fluazaindolizine between species and populations of the *Meloidogyne* spp., [155]. It does not present ovicidal activity at concentrations lower than 400 mg/L on *Meloidogyne* spp.; however, at concentrations greater than 500 mg/L, an 82% reduction in hatching of *M. incognita* eggs was observed after 3 days of exposure [156].

Greenhouse and field trials on tomato, strawberry, carrot, etc. plants show that fluazaindolizine is an effective nematicide for the treatment of *M. incognita* [157, 158]. Depending on the type of crop, application method, and soil type, the recommended application rate is between 0.25 and 2.0 kg/ha. In tomato fields infested by *M. javanica*, the application of 1.12 kg/ha of fluazaindolizine significantly reduced the number of galls for up to 2 months after planting. However, there was no change in yield or nematode population when compared to the control [159]. Similar results were obtained at concentrations lower than 1.0 kg/ha [100, 160]. Other studies indicate that at concentrations between 1.0 and 2.0 kg/ha fluazaindolizine does not reduce, after 30 days of application, the populations of *Belonolaimus longicaudatus*, *P. penetrans,* and *M. hapla*. On the other hand, the nematicidal power of flueazaindolizin e on other PPN species, as well as the influence of environmental factors on its activity, remains to be investigated.

Although the selective effect of fluazaindolizine on non-target organisms is known, there is still toxicological information on the safety of this new nematicide on human and environmental health. The LD_{50} in rats via oral administration is 1187 mg/kg [100]. The main target organs in mammals are the urinary tract, hematopoietic tissue, and liver or gallbladder [161]. It is carcinogenic in high concentrations and has little impact on the reproductive system of rats and rabbits. Despite the possible toxicological effects of fluazaindolizine for mammals, it is known to be an easily photodegradable compound. The pH does not affect its degradation in soil; however, humic acid, iron (Fe-III), and nitrate (NO_3^-) ions negatively affect the photodegradation of fluazaindolizine. It has been proposed that the possible mechanism of photodegradation of fluazaindolizine involves imidazole-ring, dechlorination, hydroxyl substitution, ring-opening, cleavage, oxidation, and decarboxylation [162, 163].

5. Conclusion

Because of their cosmopolitan nature and high reproductive success, PPNs are a problem for medium- and large-scale agriculture. Although the leading countries in high-precision agriculture have the technical and economic resources to deal with this problem, this is not the case in developing countries, where traditional pest control methods are insufficient to adequately manage PPNs. Understanding the molecular basis of the plant-nematode interaction is essential for proper control, as this will allow the development of more rational and specific methods for PPN control. Omics and *in silico* approaches based on the integrated analysis of genes and proteins involved in pathogenesis activities (invasion, feeding, colonization, etc) are key factors for the development of new, more selective, and cost-effective nematicides. Finally, organic and ecologically friendly agriculture requires the integration of new technologies with traditional control and management methods in order to reduce the impact on non-target organisms.

Acknowledgements

The author Acknowledgments Chemaxon Ltd. for the license granted for the use of the MarvinSketch software (v.22.18) with which the chemical structures of new nematicides were designed.

Conflict of interest

The author declares no conflict of interest.

Author details

Arley Rey Páez
Industrial University of Santander UIS, Bucaramanga, Colombia

*Address all correspondence to: arley.rey0914@gmail.com

References

[1] UNDESA. Department of Economic and Social Affairs. World Population Prospects 2022: Summary of Results Ten Key Messages [Internet]. 2022. Available from: https://www.un.org/development/desa/pd/sites/www.un.org.development.desa.pd/files/undesa_pd_2022_wpp_key-messages.pdf [Accessed: June 01, 2023]

[2] Andree BPJ, Chamorro A, Kraay A, Spencer P, Wang D. Predicting Food Crises. World Bank, Washington, DC; 2020. Available from: http://hdl.handle.net/10986/34510 [Accessed: 2023-06-01]

[3] FAO. Thinking about the Future of Food Safety – A Foresight Report [Internet]. Rome, Italy. 2022. Available from: 10.4060/cb8667en [Accessed: 2021-06-01]

[4] Ireland KB, van Klinken R, Cook DC, Logan D, Jamieson L, Tyson JL, et al. Plant pest impact metric system (PPIMS): Framework and guidelines for a common set of metrics to classify and prioritise plant pests. Crop Protection. 2020;**128**:105003. DOI: 10.1016/j.cropro.2019.105003

[5] FAO. Climate Change Fans Spread of Pests and Threatens Plants and Crops, New FAO Study [Internet]. 2021. Available from: https://www.fao.org/news/story/en/item/1402920/icode/ [Accessed: June 01, 2021]

[6] Lindsey APJ, Murugan S, Renitta RE. Microbial disease management in agriculture: Current status and future prospects. Biocatalysis Agricultural Biotechnology. 2020;**23**:1-12. DOI: 10.1016/j.bcab.2019.101468

[7] Patil BL. Plant viral diseases: Economic implications. In: Bamford DH, Zuckerman M, editors. Encyclopedia of Virology. 4nd ed. USA: Academic Press; 2020. pp. 81-97. DOI: 10.1016/B978-0-12-809633-8.21307-1

[8] Pan L, Li X-Z, Sun D-A, Jin H, Guo H-R, Qin B. Design and synthesis of novel coumarin analogs and their nematicidal activity against five phytonematodes. Chinese Chemical Letters. 2016;**27**:375-376. DOI: 10.1016/j.cclet.2016.01.029

[9] Singh S, Singh B, Singh AP. Nematodes: A threat to sustainability of agriculture. Procedia Environmental Sciences. 2015;**29**:215-216. DOI: 10.1016/j.proenv.2015.07.270

[10] Karuri H. Root and soil health management approaches for control of plant-parasitic nematodes in sub-Saharan Africa. Crop Protection. 2022;**152**:1-9. DOI: 10.1016/j.cropro.2021.105841

[11] Abd-Elgawad MM, Askary TH. Impact of phytonematodes on agriculture economy. In: Askary TH, Martinelli PR, editors. Biocontrol Agents of Phytonematodes. 1st ed. Washington: CABI; 2015. pp. 3-49. DOI: 10.1079/9781780643755.0003

[12] Colussi J, Schnitkey G. New Soybean Record: Historical growing of production in Brazil [Internet]. 2021. Available from: https://farmdocdaily.illinois.edu/wpcontent/uploads/2021/03/fdd300321.pdf [Accessed: June 01, 2021]

[13] Moura F, da Silva EH, Silva Antolin LA, Zanon AJ, Soares Andrade A, Antunes de Souza H, et al. Impact assessment of soybean yield and water productivity in Brazil due to climate change. European Journal of Agronomy. 2021;**129**:1-11. DOI: 10.1016/j.eja.2021.126329

[14] Tylka GL, Marett CC. Known distribution of the soybean cyst nematode, *Heterodera glycines*, in the United States and Canada in 2020. Plant Health Progress. 2021;**22**:72-74. DOI: 10.1094/PHP-10-20-0094-BR

[15] Ganguly RK, Mukherjee A, Chakraborty SK, Verma JP. Impact of agrochemical application in sustainable agriculture. In: Verma JP, Macdonald CA, Gupta VK, Podile AR, editors. New and Future Developments in Microbial Biotechnology and Bioengineering. 1st ed. Uttar Pradesh: Elsevier B.V; 2021. pp. 15-24. DOI: 10.1016/B978-0-444-64325-4.00002-X

[16] Khanna K, Kohli SK, Ohri P, Bhardwaj R. Plants-nematodes-microbes crosstalk within soil: A trade-off among friends or foes. Microbiological Research. 2021;**248**:1-23. DOI: 10.1016/j.micres.2021.126755

[17] Ahuja A, Somvanshi VS. Diagnosis of plant-parasitic nematodes using loop-mediated isothermal amplification (LAMP): A review. Crop Protection. 2020;**147**:1-10. DOI: 10.1016/j.cropro.2020.105459

[18] Ali MA, Azeem F, Li H, Bohlmann H. Smart parasitic nematodes use multifaceted strategies to parasitize plants. Frontiers in Plant Science. 2017;**8**:1699. DOI: 10.3389/fpls.2017.01699

[19] Hammond-Kosack KE, Jones JDG. Responses to plant pathogens. In: Buchanan B, Gruissem W, Jones RL, editors. Biochemistry and Molecular Biology of Plants. 2nd ed. Chichester: John Wiley & Sons; 2015. pp. 984-1049

[20] Ibrahim HMM, Ahmad EM, Martínez-Medina A, Aly MAM. Effective approaches to study the plant-root knot nematode interaction. Plant Physiology and Biochemistry. 2019;**141**:332-342. DOI: 10.1016/j.plaphy.2019.06.009

[21] Mateille T, Tavoillot J, Goillon C, Pares L, Lefèvre A, Védie H, et al. Competitive interactions in plant-parasitic nematode communities affecting organic vegetable cropping systems. Crop Protection. 2020;**135**(105206):105206. DOI: 10.1016/j.cropro.2020.105206

[22] Jones JT, Haegeman A, Danchin EGJ, Gaur HS, Helder J, Jones MGK, et al. Top 10 plant-parasitic nematodes in molecular plant pathology. Molecular Plant Pathology. 2013;**14**(9):946-961. DOI: 10.1111/mpp.12057

[23] Verdejo-Lucas S, Talavera M. Root-knot nematodes on zucchini (*Cucurbita pepo* subsp. *pepo*): Pathogenicity and management. Crop Protection. 2019;**126**:104943. DOI: 10.1016/j.cropro.2019.104943

[24] Orion D, Kritzman G, Meyer SLF, Erbe EF, Chitwood DJ. A role of the gelatinous matrix in the resistance of root-knot nematode (*Meloidogyne* spp.) eggs to microorganisms. Journal of Nematology. 2001;**33**(4):203-207. DOI: PMC2620502

[25] Karssen G, Wesemael W, Moens M. Root-knot nematodes. In: Perry RN, Moens M, editors. Plant Nematology. 2nd ed. London UK: CAB International; 2013. pp. 73-83

[26] Perry RN, Wright DJ, Chitwood DJ. Reproduction, physiology and biochemistry. In: Perry RN, Moens M, editors. Plant Nematology. 2nd ed. London UK: CAB International; 2013. pp. 235-242

[27] Pline M, Dusenbery DB. Responses of plant-parasitic nematode Meloidogyne incognita to carbon dioxide determined

by video camera-computer tracking. Journal of Chemical Ecology. 1987;**13**(4):873-888. DOI: 10.1007/BF01020167

[28] Rasmann S, Hiltpold I, Ali J. The role of root-produced volatile secondary metabolites in mediating soil interactions. In: Montanaro G, Dichio B, editors. Advances in Selected Plant Physiology Aspects. 1st ed. Rijeka, Croatia: IntechOpen; 2012. pp. 270-281. DOI: 10.5772/34304

[29] Ali JG, Alborn HT, Stelinski LL. Constitutive and induced subterranean plant volatiles attract both entomopathogenic and plant parasitic nematodes. Journal of Ecology. 2010;**99**(1):26-35. DOI: 10.1111/j.1365-2745.2010.01758.x

[30] Čepulyte R, Danquah WB, Bruening G, Williamson VM. Potent attractant for root-knot nematodes in exudates from seedling root tips of two host species. Scientific Reports. 2018;**8**(1):1-10. DOI: 10.1038/s41598-018-291

[31] Dutta TK, Ganguly AK, Gaur HS. Global status of rice root-knot nematode, *Meloidogyne graminicola*. African Journal of Microbiology Research. 2012;**6**(32):6016-6021. DOI: 10.5897/AJMR12.707

[32] Tapia-Vázquez I, Montoya-Martínez AC, De los Santos-Villalobos S, Ek-Ramos MJ, Montesinos-Matías R, Martínez-Anaya C. Root-knot nematodes (*Meloidogyne* spp.) a threat to agriculture in Mexico: Biology, current control strategies, and perspectives. World Journal of Microbiology and Biotechnology. 2022;**38**(2):1-18. DOI: 10.1007/s11274-021-03211-2

[33] Eves-van den Akker S. Plant–nematode interactions. Current Opinion in Plant Biology. 2021;**62**:102035. DOI: 10.1016/j.pbi.2021.102035

[34] Khan M, Khan AU. Plant parasitic nematodes effectors and their crosstalk with defense response of host plants: A battle underground. Rhizosphere. 2021;**17**:100288. DOI: 10.1016/j.rhisph.2020.100288

[35] Truong NM, Nguyen CN, Abad P, Quentin M, Favery B. Function of root-knot nematode effectors and their targets in plant parasitism. In: Escobar C, Fenoll C, editors. Advances in Botanical Research. Vol. 73. Toledo: Elsevier Ltd; 2015. pp. 294-319

[36] Escobar C, Barcala M, Cabrera J, Fenoll C. Overview of root-knot nematodes and giant cells. In: Escobar C, Fenoll C, editors. Advances in Botanical Research. Vol. 73. Toledo: Elsevier Ltd; 2015. pp. 5-24

[37] Jones MGK, Goto DB. Root-knot nematodes and Giant cells. In: Jones J, Gheysen G, Fenoll C, editors. Genomic and Molecular Genetics of Plant-Nematode Interactions. 1st ed. Dordrecht, Netherlands: Springer Science+Business Media; 2011. pp. 83-99

[38] Gheysen G, Mitchum MG. Phytoparasitic nematode control of plant hormone pathways. Plant Physiology. 2019;**179**(4):1212-1226. DOI: 10.1104/pp.18.01067

[39] Kyndt T, Goverse A, Haegeman A, Warmerdam S, Wanjau C, Jahani M, et al. Redirection of auxin flow in *Arabidopsis thaliana* roots after infection by root-knot nematodes. Journal of Experimental Botany. 2016;**67**(15):4559-4570. DOI: 10.1093/jxb/erw230

[40] Yuan J, Bateman P, Gutierrez-Marcos J. Genetic and epigenetic control

of transfer cell development in plants. Journal of Genetics and Genomics. 2016;**43**(9):533-539. DOI: 10.1016/j.jgg.2016.08.002

[41] Dowd CD, Chronis D, Radakovic ZS, Siddique S, Schmülling T, Werner T, et al. Divergent expression of cytokinin biosynthesis, signaling and catabolism genes underlying differences in feeding sites induced by cyst and root-knot nematodes. The Plant Journal. 2017;**92**(2):211-228. DOI: 10.1111/tpj.13647

[42] Leonetti P, Zonno MC, Molinari S, Altomare C. Induction of SA-signaling pathway and ethylene biosynthesis in *Trichoderma harzianum*-treated tomato plants after infection of the root-knot nematode *Meloidogyne incognita*. Plant Cell Reports. 2017;**36**(4):621-631. DOI: 10.1007/s00299-017-2109-0

[43] Nahar K, Kyndt T, De Vleesschauwer D, Höfte M, Gheysen G. The Jasmonate pathway is a key player in systemically induced defense against root knot nematodes in rice. Plant Physiology. 2011;**157**(1):305-316. DOI: 10.1104/pp.111.177576

[44] Jaouannet M, Magliano M, Arguel MJ, Gourgues M, Evangelisti E, Abad P, et al. The root-knot nematode calreticulin mi-CRT is a key effector in plant defense suppression. Molecular Plant-Microbe Interactions. 2013;**26**(1):97-105. DOI: 10.1094/MPMI-05-12-0130-R

[45] Przybylska A, Obrępalska-Stęplowska A. Plant defense responses in monocotyledonous and dicotyledonous host plants during root-knot nematode infection. Plant and Soil. 2020;**451**:239-260. DOI: 10.1007/s11104-020-04533-0

[46] Molinari S, Fanelli E, Leonetti P. Expression of tomato salicylic acid (SA)-responsive pathogenesis-related genes in Mi-1-mediated and SA-induced resistance to root-knot nematodes. Molecular Plant Pathology. 2013;**15**(3):255-264. DOI: 10.1111/mpp.12085

[47] Manzanilla-Lopez RH, Evans K, Bridge J. Plant diseases caused by nematode. In: Chen Z, Chen S, Dickson DW, editors. Nematology - Advances and Perspectives. Vol. 2nd. Wallingford, UK: CABI Publishing; 2004. pp. 636-716

[48] Castagnone-Sereno P, Danchin EGJ, Perfus-Barbeoch L, Abad P. Diversity and evolution of root-knot nematodes, genus *Meloidogyne*: New insights from the genomic era. Annual Review of Phytopathology. 2013;**51**(1):203-220. DOI: 10.1146/annurev-phyto-082712-102300

[49] Liu QL, Thomas VP, Williamson VM. Meiotic parthenogenesis in a root-knot nematode results in rapid genomic homozygosity. Genetics. 2007;**176**(3):1483-1490. DOI: 10.1534/genetics.107.071134

[50] Chitwood DJ, Perry RN. Reproduction, physiology and biochemistry. In: Perry NR, Moens M, Starr JL, editors. Roo-Knot Nematodes. 1st ed. London, UK: CABI Publishing; 2009. pp. 182-194

[51] Baldwin JG, Handoo ZA. General morphology of cyst nematodes. In: Perry RN, Moens M, Jones JT, editors. Cyst Nematodes. 1st ed. London, UK: CABI Publishing; 2018. pp. 337-359

[52] Smiley RW, Dababat AA, Iqbal S, Jones MGK, Maafi ZT, Peng D, et al. Cereal cyst nematodes: A complex and destructive group of heterodera species. Plant Disease. 2017;**101**(10):1692-1720. DOI: 10.1094/PDIS-03-17-0355-FE

[53] Palomares-Rius JE, Hedley P, Cock PJA, Morris JA, Jones JT, Blok VC. Gene expression changes in diapause or quiescent potato cyst nematode, *Globodera pallida*, eggs after hydration or exposure to tomato root diffusate. Peer J. 2016;**4**:e1654. DOI: 10.7717/peerj.1654

[54] Shimizu K, Kushida A, Akiyama R, Lee HJ, Okamura Y, Masuda Y, et al. Hatching stimulation activity of steroidal glycoalkaloids toward the potato cyst nematode, *Globodera rostochiensis*. Plant Biotechnology. 2020;**37**(3):319-325. DOI: 10.5511/plantbiotechnology.20.0516a

[55] Ngala B, Mariette N, Ianszen M, Dewaegeneire P, Denis MC, Porte C, et al. Hatching induction of cyst nematodes in bare soils drenched with root exudates under controlled conditions. Frontiers in Plant Science. 2021;**11**:602825. DOI: 10.3389/fpls.2020.602825

[56] Shimizu K, Akiyama R, Okamura Y, Ogawa C, Masuda Y, Sakata I, et al. Solanoeclepin B, a hatching factor for potato cyst nematode. Science Advances. 2023;**9**(11):1-11. DOI: 10.1126/sciadv.adf4166

[57] Nonaka S, Katsuyama T, Kondo T, Sasaki Y, Asami T, Yajima S, et al. 1,10-phenanthroline and its derivatives are novel hatching stimulants for soybean cyst nematodes. Bioorganic and Medicinal Chemical Letters. 2016;**1**:5240-5243. DOI: 10.1016/j.bmcl.2016.09.052

[58] Perry RN. Plant signal in nematode hatching and attraction. In: Fenoll C, Grundler FMW, Ohl SA, editors. Cellular and Molecular Aspects of Plant-Nematode Interactions. Vol. 10. Dordrecht, Netherlands: Kluwer Academic Publishers; 1997. pp. 38-46

[59] Perry RN. Hatching. In: Lee DL, editor. The Biology of Nematodes. 1st ed. London, UK: Taylor & Francis Group; 2002. pp. 294-335

[60] Bohlmann H. Introductory chapter on the basic biology of cyst nematodes. In: Escobar C, Fenoll C, editors. Advances in Botanical Research. Vol. 73. Toledo: Elsevier Ltd; 2015. pp. 2-21. DOI: 10.1016/bs.abr.2014.12.001

[61] Masler EP, Perry RN. Hatch, survival and sensory perception. In: Perry RN, Moens M, Jones JT, editors. Cyst nematodes. 1st ed. London, UK: CABI International; 2018. pp. 44-73

[62] Perry RN, Clarke A, Hennessy J. The influence of osmotic pressure on the hatching of Heterodera schachtii. Revue Nematol. 1980;**3**(1):3-9

[63] Li T, Tang BZ, Li B, Mobbs D. The hatching of cereal cyst nematode (*Heterodera avenae*) in response to different inorganic ions. Scientia Agricola. 2022;**79**(1):1-7. DOI: 10.1590/1678-992X

[64] Turner SJ, Subbotin SA. Cyst nematodes. In: Perry RN, Moens M, editors. Plant nematology. 2nd ed. London, UK: CABI International; 2013. pp. 109-141

[65] Price JA, Coyne D, Blok VC, Jones JT. Potato cyst nematodes *Globodera rostochiensis* and *G. pallida*. Molecular Plant Pathology. 2021;**22**(5):495-507. DOI: 10.1111/mpp.13047

[66] Curtis RHC. Plant-nematode interactions: Environmental signals detected by the nematode's chemosensory organs control changes in the surface cuticle and behaviour. Parasite. 2008;**15**(3):310-316. DOI: 10.1051/parasite/2008153310

[67] Mende NV, Nobre MJ, Perry RN. Host finding, invasion and feeding. In: Sharma SB, editor. The Cyst Nematodes. 1st ed. Dordrecht, Netherlands: Springer Science+Business Media; 1998. pp. 217-238

[68] Sobczak M, Golinowski W. Cyst nematodes and syncytia. In: Jones J, Gheysen G, Fenoll C, editors. Genomic and Molecular Genetics of Plant-Nematode Interactions. 1st ed. Dordrecht, Netherlands: Springer Science+Business Media; 2011. pp. 61-80

[69] Wyss U. Observations on the feeding behaviour of *Heterodera schachtii* throughout development, including events during moulting. Fundamental and Applied Nematology. 1992;**15**:75-89

[70] Subbotin SA, Mundo-Ocampo M, Baldwin JG. Biology and evolution. In: Subbotin SA, Mundo-Ocampo M, Baldwin JG, editors. Systematics of Cyst Nematodes (Nematoda: Heteroderinae), Part a. 1st ed. Leiden, Netherlands: BRILL; 2010. pp. 43-70

[71] Sobczak M, Golinowski W. Structure of cyst nematode feeding sites. In: Berg RH, Taylor CG, editors. Cell Biology of Plant Nematodes Parasitism. Vol. 15. Heidelberg, Germany: Springer-Verlag; 2009. pp. 153-187

[72] Mitchum MG, Hussey RS, Baum TJ, Wang X, Elling AA, Wubben M, et al. Nematode effector proteins: An emerging paradigm of parasitism. New Phytologist. 2013;**199**(4):879-894. DOI: 10.1111/nph.12323

[73] Karczmarek A, Overmars H, Helder J, Goverse A. Feeding cell development by cyst and root-knot nematodes involves a similar early, local and transient activation of a specific auxin-inducible promoter element. Molecular Plant Pathology. 2004;**5**(4):343-346. DOI: 10.1111/j.1364-3703.2004.00230.x

[74] Jones MGK, Fosu-Nyarko J. Molecular biology of root lesion nematodes (*Pratylenchus* spp.) and their interaction with host plants. Annals of Applied Biology. 2014;**164**(2):163-181. DOI: 10.1111/aab.12105

[75] Bucki P, Qing X, Castillo P, Gamliel A, Dobrinin S, Alon T, et al. The genus *Pratylenchus* (Nematoda: Pratylenchidae) in Israel: From taxonomy to control practices. Plants. 2020;**9**(11):1475. DOI: 10.3390/plants9111475

[76] Mokrini F, Viaene N, Waeyenberge L, Dababat AA, Moens M. Root-lesion nematodes in cereal fields: Importance, distribution, identification, and management strategies. Journal of Plant Diseases and Protection. 2018;**126**(1):1-11. DOI: 10.1007/s41348-018-0195-z

[77] Orlando V, Grove IG, Edwards SG, Prior T, Roberts D, Neilson R, et al. Root-lesion nematodes of potato: Current status of diagnostics, pathogenicity and management. Plant Pathology. 2020;**69**(3):405-417. DOI: 10.1111/ppa.13144

[78] Zunke U. Ectoparasitic feeding behaviour of the root lesion nematode, *Pratylenchus penetrans*, on root hairs of different host plants. Revue Nématologie. 1990;**13**(3):331-337. DOI: 10670/1.u6a71m

[79] Ducan LW, Moens M. Migratory endoparasitic nematodes. In: Perry RN, Moens M, editors. Plant nematology. 2nd ed. London UK: CAB International; 2013. pp. 145-178

[80] Fosu-Nyarko J, Jones MP. Advances in understanding the molecular mechanisms of root lesion nematode host interactions.

Annual Review of Phytopathology. 2016;**54**(1):253-278. DOI: 10.1146/annurev-phyto-080615-100257

[81] Vieira PC, Shao J, Vijayapalani P, Maier T, Pellegrin C, Eves-Van S, et al. A new esophageal gland transcriptome reveals signatures of large scale de novo effector birth in the root lesion nematode *Pratylenchus penetrans*. BMC Genomics. 2020;**21**(1):1-16. DOI: 10.1186/s12864-020-07146-0

[82] Devi AN, Ponnuswami V, Sundararaju P, Soorianathasundaram K, Sathiamoorthy S, Uma S, et al. Mechanism of resistance in banana cultivars against root lesion nematode, *Pratylenchus coffeae*. Indian Journal of Nematology. 2007;**37**:138-144

[83] Perry RN. Desiccation survival of parasitic nematodes. Parasitology. 1999;**119**:S19-S30. DOI: 10.1017/S0031182000084626

[84] Thompson JP, Rostad HE, Whish JPM. Survival of root-lesion nematodes (*Pratylenchus thornei*) after wheat growth in a vertisol is influenced by rate of progressive soil desiccation. Annals of Applied Biology. 2016;**170**(1):78-88. DOI: 10.1111/aab.12316

[85] Ribeiro LM, Campos HD, Neves DL, Dias-Arieira CR. Survival of *Pratylenchus brachyurus* under dry soil conditions. Heliyon. 2020;**6**(9):1-6. DOI: 10.1016/j.heliyon.2020.e05075

[86] Townshend JL. Anhydrobiosis in *Pratylenchus penetrans*. Journal of Nematology. 1984;**16**(3):282-289. DOI: PMC2618384/?tool=EBI

[87] Storey RMJ, Glazer I, Orion D. Lipid utilisation by starved and anhydrobiotic individuals of *Pratylenchus thornei*. Nematologica. 1982;**28**(4):373-378. DOI: 10.1163/187529282X00178

[88] FAO. FAOSTAT [Internet]. www.fao.org. 2022. Available from: https://www.fao.org/faostat/en/#data/QCL

[89] Favoreto L, Meyer MC, Dias-Arieira CR, Machado AC, Santiago DC, Ribeiro N. Diagnose e manejo de fitonematoides na cultura da soja. Informe Agropecuario. 2019;**306**:18-29

[90] Pires D, Vicente CS, Menéndez E, Faria JMS, Rusinque L, Camacho MJ, et al. The fight against plant-parasitic nematodes: Current status of bacterial and fungal biocontrol agents. Pathogens. 2022;**11**(10):1-22. DOI: 10.3390/pathogens11101178

[91] Rajak P, Roy S, Ganguly A, Mandi M, Dutta A, Das K, et al. Agricultural pesticides – Friends or foes to biosphere? Journal of Hazardous Materials Advances. 2023;**10**:100264. DOI: 10.1016/j.hazadv.2023.100264

[92] Zasada IA, Halbrendt JM, Kokalis-Burelle N, LaMondia J, McKenry MV, Noling JW. Managing nematodes without methyl bromide. Annual Review of Phytopathology. 2010;**48**(1):311-328. DOI: 10.1146/annurev-phyto-073009-114425

[93] Sharma A, Kumar V, Shahzad B, Tanveer M, Sidhu GPS, Handa N, et al. Worldwide pesticide usage and its impacts on ecosystem. SN Applied Sciences. 2019;**1**(11):11446. DOI: g/10.1007/s42452-019-1485-1

[94] Wram CL, Zasada IA. Short-term effects of sublethal doses of nematicides on *Meloidogyne incognita*. Phytopathology. 2019;**109**(9):1605-1613. DOI: 10.1094/PHYTO-11-18-0420-R

[95] Spherical Insights LLP. Global Nematicides Market Size to Worth USD 4.28 Billion By 2032 | CAGR of 9.1% [Internet]. Available from:

https://www.sphericalinsights.com/request-sample/1768 . [Accessed: May 21, 2023]

[96] Taylor RWD. Methyl bromide—Is there any future for this noteworthy fumigant? Journal of Stored Products Research. 1994;**30**(4):253-260. DOI: 10.1016/S0022-474X(94)90317-4

[97] Sharma A, Shukla A, Attri K, Kumar M, Kumar P, Suttee A, et al. Global trends in pesticides: A looming threat and viable alternatives. Ecotoxicology and Environmental Safety. 2020;**201**:110812. DOI: 10.1016/j.ecoenv.2020.110812

[98] Costa LG, Giordano G, Guizzetti M, Vitalone A. Neurotoxicity of pesticides: A brief review. Frontiers in Bioscience. 2008;**13**(13):1240. DOI: 10.2741/2758

[99] Gupta RC, Sachana M, Mukherjee IM, Doss RB, Malik JK, Milatovic D. Organophosphates and carbamates. In: Gupta RC, editor. Veterinary Toxicology. 3nd ed. London: Academic Press; 2018. pp. 495-507. DOI: 10.1016/B978-0-12-811410-0.00037-4

[100] Oka Y. From old-generation to next-generation nematicides. Agronomy. 2020;**10**(9):1387. DOI: 10.3390/agronomy10091387

[101] Sparks TC, Crossthwaite AJ, Nauen R, Banba S, Cordova D, Earley F, et al. Insecticides, biologics and nematicides: Updates to IRAC's mode of action classification - a tool for resistance management. Pesticide Biochemistry and Physiology. 2020;**167**:104587. DOI: 10.1016/j.pestbp.2020.104587

[102] Grange R, Mandiriza G, Zyl C. Nematodes, nematicides and resistance management. In: Insecticide Resistence Action Committee. 2021; South Africa. p. 1-23. Available from: https://irac-online.org/documents/nematicides-and-resistance-management/ [Accessed:2023-05-21]

[103] Faske TR, Hurd K. Sensitivity of Meloidogyne incognita and Rotylenchulus reniformis to Fluopyram. Journal of Nematology. 2015;**47**(4):316-321

[104] Desaeger J, Wram C, Zasada I. New reduced-risk agricultural nematicides - rationale and review. Journal of Nematology. 2020;**52**(1):1-16. DOI: 10.21307/jofnem-2020-091

[105] Xue Z, Zhong S, Shen J, Sun Y, Gao X, Wang X, et al. Multiple mutations in SDHB and SDHC2 subunits confer resistance to the succinate dehydrogenase inhibitor cyclobutrifluram in *Fusarium fujikuroi*. Journal of Agricultural and Food Chemistry. 2023;**71**(8):3694-3704. DOI: 10.1021/acs.jafc.2c08023

[106] Liang P, Shen S, Xu Q, Zhang L, Jin S, Lu H, et al. Design, synthesis biological activity, and docking of novel fluopyram derivatives containing guanidine group. Bioorganic and Medical Chemistry. 2021;**29**:115846-115846. DOI: 10.1016/j.bmc.2020.115846

[107] Harada S, Inaoka DK, Ohmori J, Kita K. Diversity of parasite complex II. Biochimica et Biophysica Acta (BBA) – Bioenergetics. 2013;**1827**(5):658-667. DOI: 10.1016/j.bbabio.2013.01.005

[108] Kuramochi T, Hirawake H, Kojima S, Takamiya S, Furushima R, Aoki T, et al. Sequence comparison between the flavoprotein subunit of the fumarate reductase (complex II) of the anaerobic parasitic nematode, *Ascaris suum* and the succinate dehydrogenase of the aerobic, free-living nematode, *Caenorhabditis*

elegans. Molecular and Biochemical Parasitology. 1994;**68**(2):177-187. DOI: 10.1016/0166-6851(94)90163-5

[109] Karavaeva V, Sousa FL. Modular structure of complex II: An evolutionary perspective. Biochimica et Biophysica Acta (BBA) - Bioenergetics. 2022;**1864**:1-11. DOI: 10.1016/j.bbabio.2022.148916

[110] Lümmen P, Fürsch H. Fluopyram a novel nematicide for the control of root-knot nematodes. In: Jeschke P, Witschel M, Krämer W, Schirmer U, editors. Modern Crop Protection Compounds. 3nd ed. Weinheim: Wiley-VCH Verlag GmbH & Co. KGaA; 2019. pp. 1630-1640

[111] Burns AR, Luciani GM, Musso G, Bagg R, Yeo M, Zhang Y, et al. *Caenorhabditis elegans* is a useful model for anthelmintic discovery. Nature Communications. 2015;**6**(1):1-11. DOI: 10.1038/ncomms8485

[112] Oka Y, Saroya Y. Effect of fluensulfone and fluopyram on the mobility and infection of second-stage juveniles of *Meloidogyne incognita* and *M. javanica*. Pest Management Science. 2019;**75**(8):2095-2106. DOI: 10.1002/ps.5399

[113] Schleker ASS, Rist M, Matera C, Damijonaitis A, Collienne U, Matsuoka K, et al. Mode of action of fluopyram in plant-parasitic nematodes. Scientific Reports. 2022;**12**(1):1-14. DOI: 10.1038/s41598-022-15782-7

[114] Storelli A, Keiser A, Eder R, Jenni S, Kiewnick S. Evaluation of fluopyram for the control of *Ditylenchus dipsaci* in sugar beet. Journal of Nematology. 2020;**52**(1):1-10. DOI: 10.21307/jofnem-2020-071

[115] Feist E, Kearn J, Gaihre Y, O'Connor V, Holden-Dye L. The distinct profiles of the inhibitory effects of fluensulfone, abamectin, aldicarb and fluopyram on *Globodera pallida* hatching. Pesticide Biochemistry and Physiology. 2020;**165**:104541. DOI: 10.1016/j.pestbp.2020.02.007

[116] Hawk T, Faske TR. Effect of seed-applied fluopyram on *Meloidogyne incognita* infection and maturity in cotton and soybean. Journal of Nematology. 2021;**52**:1-7. DOI: 10.21307/jofnem-2020-083

[117] Grabau ZJ, Liu C, Schumacher LA, Small IM, Wright DL. In-furrow fluopyram nematicide efficacy for *Rotylenchulus reniformis* management in cotton production. Crop Protection. 2021;**140**:105423. DOI: 10.1016/j.cropro.2020.105423

[118] Waisen P, Wang KH, Uyeda J, Myers RY. Effects of fluopyram and azadirachtin integration with sunn hemp on nematode communities in zucchini, tomato and sweet potato in Hawaii. Journal of Nematology. 2021;**53**(1):1-15. DOI: 10.21307/jofnem-2021-030

[119] Jeanmart S, Edmunds A, Lamberth C, Pouliot M. Synthetic approaches to the 2010-2014 new agrochemicals. Bioorganic and Medical Chemistry. 2016;**24**(3):317-341. DOI: 10.1016/j.bmc.2015.12.014

[120] USEPA. Pesticide Product Label, FLUENSULFONE 480EC [Internet]. 2014. Available from: https://www3.epa.gov/pesticides/chem_search/ppls/066222-00243-20140911.pdf [Accessed: June 01, 2023]

[121] Dewhurst I, Tasheva M. Fluensulfone. In: WHO. Pesticide Residues in Food – 2013: Toxicological Evaluations. 1st ed. Geneva: World Health Organization; 2015. pp. 271-309

[122] Proença P, Teixeira H, De Mendonça MC, Castanheira F, Marques EP, Corte-Real F, et al. Aldicarb poisoning: One case report. Forensic Science International. 2004;**146**:S79-S81. DOI: 10.1016/j.forsciint.2004.09.028

[123] APVMA. Public Release Summery on the Evaluation of the New Active Fluensulfone in the Product Nimitz 480 EC Nematicide; Australian Pesticides and Veterinary Medicines Authority: Kingstone, Australia, 2015; p. 47. Available from: http://apvma.gov.au/node/15596 [Accessed: 2023-06-01]

[124] Giannakou IO, Panopoulou S. The use of fluensulfone for the control of root-knot nematodes in greenhouse cultivated crops: Efficacy and phytotoxicity effects. Cogent Food & Agriculture. 2019;**5**(1):1643819. DOI: 10.1080/23311932.2019.1643819

[125] Kearn J, Ludlow E, Dillon J, O'Connor V, Holden-Dye L. Fluensulfone is a nematicide with a mode of action distinct from anticholinesterases and macrocyclic lactones. Pesticide Biochemistry and Physiology. 2014;**109**:44-57. DOI: 10.1016/j.pestbp.2014.01.004

[126] Kawanobe T, Fujita H. Evaluation of Nematicidal activity of Fluensulfone against non-target free-living nematodes under field conditions. Agronomy. 2019;**9**(12):853. DOI: 10.3390/agronomy9120853

[127] Waldo BD, Grabau ZJ, Mengistu TM, Crow WT. Nematicide effects on non-target nematodes in bermudagrass. Journal of Nematology. 2019;**51**:e2019-e2009. DOI: 10.21307/jofnem-2019-009

[128] Oka Y, Shuker S, Tkachi N. Nematicidal efficacy of MCW-2, a new nematicide of the fluoroalkenyl group, against the root-knot nematode *Meloidogyne javanica*. Pest Management Science. 2009;**65**(10):1082-1089. DOI: 10.1002/ps.1796

[129] Maienfisch P, Loiseleur O, Slaats B. Recent nematicides. In: Jeschke P, Witschel M, Krämer W, Schirmer U, editors. Modern Crop Protection Compounds. 3nd ed. Weinheim: Wiley-VCH Verlag GmbH & Co. KGaA; 2019. pp. 1630-1640

[130] Pitterna T, Böger M, Maienfisch P. Gem-difluorovinyl derivatives as insecticides and acaricides. Chimia. 2004;**58**(3):108-116. DOI: 10.2533/000942904777678163

[131] Oka Y. Sensitivity to fluensulfone of inactivated *Meloidogyne* spp. second-stage juveniles. Pest Management Science. 2020;**76**(7):2379-2387. DOI: 10.1002/ps.5774

[132] Oka Y. Nematicidal activity of fluensulfone against some migratory nematodes under laboratory conditions. Pest Management Science. 2014;**70**(12):1850-1858. DOI: 10.1002/ps.3730

[133] Grabau ZJ, Noling JW, Navia Gine PA. Fluensulfone and 1,3-dichloroprene for plant-parasitic nematode management in potato production. Journal of Nematology. 2019;**51**:1-12. DOI: 10.21307/jofnem-2019-038

[134] Ludlow K. Public Release Summary on the Evaluation of the New Active Fluensulfone in Product NIMITZ 480 EC Nematicide. Australian Pesticides and Veterinary Medicines Authority, Kingston [Internet]. 2015. Available from: https://apvma.gov.au/sites/default/files/publication/15596-prs-fluensulfone.pdf [Accessed: June 01, 2023]

[135] Norshie PM, Grove IG, Back MA. Persistence of the nematicide fluensulfone in potato (*Solanum tuberosum* ssp. tuberosum) beds under field conditions. Nematology. 2017;**19**(6):739-747. DOI: 10.1163/15685411-00003085

[136] Faske T, Brown KA, Kelly JC. Toxicity of Tioxazafen to *Meloidogyne incognita* and *Rotylenchulus reniformis*. Journal of Nemtology. 2022;**54**(1):1-6. DOI: 10.2478/jofnem-2022-0007

[137] Slomczynska U, South MS, Bunkers GJ, Edgecomb D, Wyse-Pester D, Selness S, et al. Tioxazafen: A new broad-spectrum seed treatment nematicide. In: ACS Symposium Series. Washington, DC: American Chemical Society; 2015. pp. 129-147. DOI: 10.1021/bk-2015-1204.ch010

[138] South SM, Wilson D, Spal S, Slomczynska U, Bunkers GJ, et al. Development of Tioxazafen as a new broad spectrum Nematicide. In: Jeschke P, Witschel M, Kramer W, Schirmer U, editors. Modern Crop Protection Compounds. 1st ed. Germany: Wiley-VCH; 2019. pp. 1615-1628. DOI: 10.1002/9783527644179

[139] Gopalsamy A, Shi M, Golas J, Vogan E, Jacob J, Johnson M, et al. Discovery of benzisoxazoles as potent inhibitors of chaperone heat shock protein 90. Journal Medical Chemistry. 2008;**51**(3):373-375. DOI: 10.1021/jm701385c

[140] Lamberth C. Oxazole and isoxazole chemistry in crop protection. Journal of Heterocyclic Chemistry. 2018;**55**(9):2035-2045. DOI: 10.1002/jhet.3252

[141] Wayne N, Mishra P, Bolon DN. Hsp90 and client protein maturation. In: Calderwood SK, Prince TL, editors. Molecular Chaperones: Methods in Molecular Biology. Vol. 787. New York: Springer Science+Business Media; 2011. pp. 33-44. DOI: 10.1007/978-1-61779-295-3_3

[142] Devaney E, O'Neill K, Harnett W, Whitesell L, Kinnaird JH. Hsp90 is essential in the filarial nematode *Brugia pahangi*. International Journal for Parasitology. 2005;**35**(6):627-636. DOI: 10.1016/j.ijpara.2005.01.007

[143] Gelmedin V, Delaney A, Jennelle L, Hawdon JM. Expression profile of heat shock response factors during hookworm larval activation and parasitic development. Molecular and Biochemical Parasitology. 2015;**202**(1):1-14. DOI: 10.1016/j.molbiopara.2015.08.003

[144] Somogyvári M, Gecse E, Sőti C. DAF-21/Hsp90 is required for *C. elegans* longevity by ensuring DAF-16/FOXO isoform a function. Scientific Reports. 2018;**8**(1):12048. DOI: 10.1038/s41598-018-30592-6

[145] Wolterink G, Boobis AR. Tioxazafen. In: Joint Meeting of the FAO Panel of Experts on Pesticide Residues in Food and the Environment and the WHO Core Assessment Group on Pesticide Residues. Pesticide Residues in Food - 2018: Toxicological Evaluations. Berlin, Germany, 18-27 September 2018. Geneva: World Health Organization and Food and Agriculture Organization of the United Nations; 2019. Licence: CC BY-NC-SA 3.0 IGO. Available from: https://inchem.org/documents/jmpr/jmpmono/v2018pr01.pdf [Accessed: 2023-06-01]

[146] Uzuele EL. Eficácia do tratamento de sementes com o nematicida tioxazafen no controle dos nematoides *Heterodera glycines*, *Meloidogyne incognita*, M. javanica, Pratylenchus brachyurus e P. zeae [thesis]. Piracicaba: Universidade de São Paulo; 2016

[147] Walker DM, Graham CR, Miller WR, Koeller KJ. Three step, one-pot process to prepare thiophene-2-carbonyl chloride (TCC), a key raw material in the manufacture of tioxazafen (Nemastrike®). Tetrahedron Letters. 2019;**60**(12):834-838. DOI: 10.1016/j.tetlet.2019.02.015

[148] Lahm GP, Wiles JA, Cordova D, Thoden T, Desaeger J, Smith BK. Fluazaindolizine: A new active ingredient for the control of plant-parasitic nematodes. In: Jeschke P, Witschel M, Krämer W, Schirmer U, editors. Modern Crop Protection Compounds. 3nd ed. Weinheim: Wiley-VCH Verlag GmbH & Co. KGaA; 2019. pp. 1630-1640

[149] EPA. Memorandum Supporting Revised Proposed Decision to Approve Registration for the New Active Ingredient Fluazaindolizine [Internet]. 2023. Available from: https://www.regulations.gov/search?filter=EPA-HQ-OPP-2020-0065 [Accessed: June 04, 2023]

[150] Groover WL, Lawrence KS. Evaluation of a new chemical nematicide, fluazaindolizine (Reklemel™ active), for plant-parasitic nematode management in bermudagrass. Journal of Nematology. 2021;**53**:1-14. DOI: 10.21307/jofnem-2021-043

[151] Leonardi R, Jackowski S. Biosynthesis of pantothenic acid and coenzyme a. EcoSal Plus. 2007;**2**(2):1-22. DOI: 10.1128/ecosalplus.3.6.3.4

[152] Craig JP, Bekal S, Niblack T, Domier L, Lambert KN. Evidence for horizontally transferred genes involved in the biosynthesis of vitamin B(1), B(5), and B(7) in *Heterodera glycines*. Journal of Nematology. 2009;**41**(4):281-290

[153] Siddique S, Radakovic ZS, Hiltl C, Pellegrin C, Baum T, Beasley H, et al. The genome and lifestage-specific transcriptomes of a plant-parasitic nematode and its host reveal susceptibility genes involved in trans-kingdom synthesis of vitamin B5. Nature Communications. 2022;**13**(1):6190. DOI: 10.1038/s41467-022-33769-w

[154] Thoden TC, Wiles JA. Biological attributes of Salibro™, a novel sulfonamide nematicide. Part 1: Impact on the fitness of *Meloidogyne incognita*, *M. hapla* and *Acrobeloides buetschlii*. Nematology. 2019;**21**(6):625-639

[155] Thoden T, Pardavella IV, Tzortzakakis EA. In vitro sensitivity of different populations of *Meloidogyne javanica* and *M. incognita* to the nematicides Salibro™ and Vydate®. Nematology. 2019;**21**(8):889-893

[156] Wu HY, de Oliveira SJ, Becker JS, Becker JO. Fluazaindolizine mitigates plant-parasitic nematode activity at sublethal dosages. Journal of Pest Science. 2020;**94**(2):573-583. DOI: 10.1007/s10340-020-01262-2

[157] Qiao K, Liu Q, Zhang S. Evaluation of fluazaindolizine, a new nematicide for management of Meloidogyne incognita in squash in calcareous soils. Crop Protection. 2021;**143**:105469. DOI: 10.1016/j.cropro.2020.105469

[158] Watson TT, Desaeger JA. Evaluation of non-fumigant chemical and biological nematicides for strawberry production in Florida. Crop Protection. 2019;**117**:100-107. DOI: 10.1016/j.cropro.2018.11.019

[159] Desaeger JA, Watson TT. Evaluation of new chemical and biological nematicides for managing *Meloidogyne javanica* in tomato production and associated double-crops in Florida. Pest Management Science. 2019;**75**(12):3363-3370. DOI: 10.1002/ps.5481

[160] Regmi H, Desaeger J. Integrated management of root-knot nematode (*Meloidogyne* spp.) in Florida tomatoes combining host resistance and nematicides. Crop Protection. 2020;**134**:105170. DOI: 10.1016/j.cropro.2020.105170

[161] USEPA. Memorandum Supporting Revised Proposed Decision to Approve Registration for the New Active Ingredient Fluazaindolizine [Internet]. 2023. Available from: https://www.regulations.gov/search?filter=EPA-HQ-OPP-2020-0065

[162] Lin H, Pang K, Ma Y, Hu J. Photodegradation of fluazaindolizine in water under simulated sunlight irradiation: Identification of transformation products and elucidation of transformation mechanism. Chemosphere. 2019;**214**:543-552. DOI: 10.1016/j.chemosphere.2018.09.151

[163] Pang N, Lin H, Hu J. Photodegradation of fluazaindolizine in aqueous solution with graphitic carbon nitride nanosheets under simulated sunlight illumination. Ecotoxicology and Environmental Safety. 2019;**170**:33-38. DOI: 10.1016/j.ecoenv.2018.11.019

Chapter 4

Use of Plant Material in the Management of Plant Parasitic Nematodes

Mohammed Bukar Aji

Abstract

Use of synthetic chemicals creates significant environmental dangers, which is a significant global issue. As a result, researchers have looked into plant materials and taken on the challenge of finding more environmentally friendly alternatives. As organic soil amendments, dried leaves, seed powders and cakes, tree fibers, and green manures all possess nematicidal qualities. It is permitted to combine them with other cultural techniques as an organic soil addition. They might be extracted chemically using acetone, methanol, or ethanol, or they could be prepared as aqueous extracts, root extracts, or exudates for use as extracts. They were used as a soil drench, a root dip, and as a foliar application. The plant materials have a significant impact on altering the soil's ecosystem, and if properly developed, they could result in the long-awaited ecological alternatives to synthetic nematicides. It would be cheap for the resource-strapped subsistence farmer to control plant parasitic nematodes by utilizing botanical nematicides.

Keywords: plant parasitic nematodes, medicinal plants, bare-root dip, soil drench, soil amendment

1. Introduction

Nematodes are basic animals often only containing 1000 cells or less [1]. Nematodes throughout or part of their life cycle are worm-shaped, also called vermiform, even though some species become enlarged and rounded in later life stages [2]. According to Lambert and Bekal [1], a nematode's basic body structure is a tube inside of a tube. Their outer skin, or cuticle, is produced by the inner hypodermis. The muscles can only move dorsally and ventrally since they are longitudinally linked to the nematode's hypodermis. An inner tube, the alimentary canal, and fluid that applies pressure to the body wall to preserve shape and permit movement are all inside the nematode. A hollow mouth spear known as a stylet is located at the head of a plant parasitic worm (**Figure 1**). According to Sato et al. [4], the worm uses this stylet to puncture plant cells, absorb food, and create proteins and metabolites that aid the nematode in parasitizing the plant. The stylet is connected to the pharynx, which is subsequently connected to the intestinal ends, by the rectum in the case of

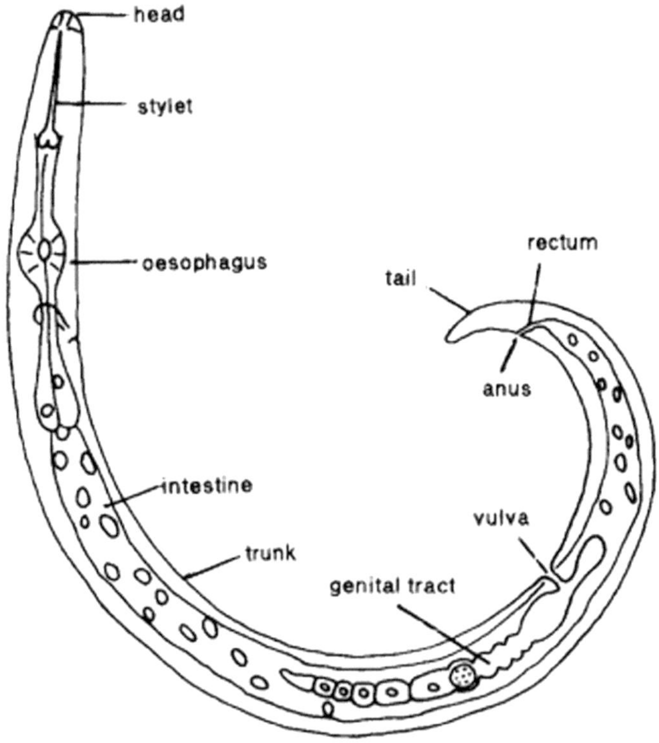

Figure 1.
A typical nematode structure (Source: [3]).

the female nematode and the cloaca in the case of the male. According to Lambert and Bekal [1], three to five salivary glands that are connected to the pharynx produce secretions that may be discharged from the stylet and aid the nematode in invading and parasitizing plants.

Plant parasitic nematodes pose a severe threat to crop yield globally [5, 6]. Nicol et al. [7] predicted that they might lead to yearly losses of up to 25%, or more than 80 billion US dollars. The primary cause of these losses is an endoparasitic worm that is notoriously difficult to control and that resides and feeds inside the root tissue [4]. The Solanaceae family (potato, tomato, and pepper), Gramineae (rice, wheat, and maize), Malvaceae (cotton, okra), and Fabaceae (soybean, cowpea), among others, are among the host plants that plant parasitic nematodes feed on [4].

Because of the significant nematode population decline and yield gain brought about by the use of chemical nematicides, plant parasitic nematode control/management (**Figure 2**) has relied extensively on them for decades [8, 9]. Chemical nematicides are the most efficient and quickly acting nematode management techniques, but they are also dangerous to the environment and people's health [10]. Nematicides, like methyl bromide, are poisonous if swallowed or absorbed through skin contact [11]. A central nervous system depression and allergic disease may result from repeated exposure to chemical nematicides [12]. Additionally, they are relatively pricey for many small farmers. Alternative management strategies for plant parasitic nematodes that do not harm the environment have received a lot of attention from researchers [13]. This chapter focuses on environmentally acceptable management solutions for plant parasitic nematodes employing plant material

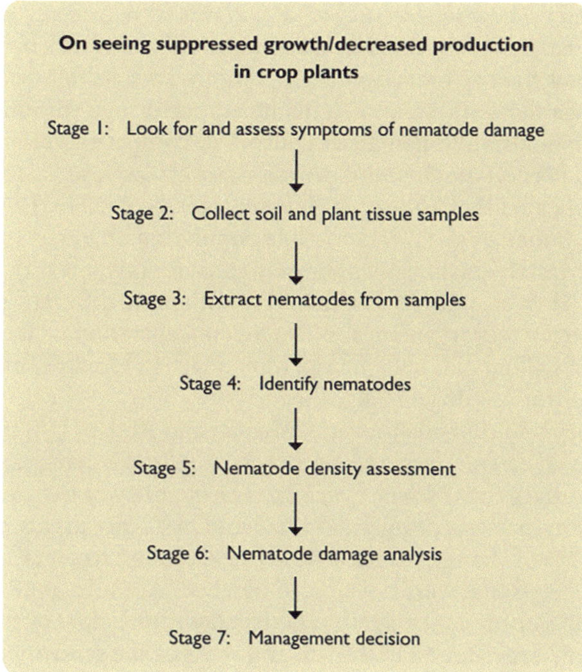

Figure 2.
The stages in nematode assessment and management.

(powder and extracts) and testing the effectiveness of various treatment techniques (bare-root dip, soil soaking, and amendment).

2. Use of plant extract in control plant parasitic nematode

Ononuju and Nzenwa [14] investigated the impact of cold and hot aqueous extracts of five plant specimens (*Luffa cylindrica*, *Momordica charantia*, *Euphorbia hirta*, *Desmodium scorpiurus*, and *Stachytarpheta cayennensis*), the wood ash of *Gmelina arborea*, a synthetic insecticide (Karate-Lambda cyahalothrin), and untreated tap water (control) on the egg hatchability as well as control of *Meloidogyne* spp. in cowpea (*Vigna unguiculata* (L.) Walp in the laboratory and greenhouse. The ability of nematode eggs to hatch was significantly decreased by *L. cylindrica* hot water extract (HWE) and *M. charantia* cold water extract (CWE), according to laboratory results. According to greenhouse results, the cold water extract (CWE) of *S. cayennensis*, the HWE of *L. cylindrica*, and the HWE of *E. hirta* all considerably increased cowpea yield. The populations of nematodes in the soil, on the roots, and in the galls on the roots were reduced by the aqueous extracts. The outcome of *G. arborea* extracts and synthesized Karate-Lambda-cyhalothrin did not differ noticeably. Umar and Mohammed [15] described the impact of water hyacinth leaf extracts on *Meloidogyne incognita* juvenile mortality in a laboratory setting. Young *M. incognita* were exposed to the leaf's crude extract and diluted extracts for 120 h. Juvenile mortality was 100% in the crude extracts. Additionally, they noted that as exposure time increased, juvenile mortality did as well. In order to suppress *M. incognita* on soybeans, Umar and Aji [16] investigated the effects of two organic amendments, namely bitter leaf (BL) and cashew seed kernel (CSK). Plants were infected at the base with 1000 juvenile *M. incognita* 3 weeks after germination. Each plant's base soil was mixed with a bag of each of the two amendments separately 2 weeks following germination. The study's findings showed that treated plants outperformed the untreated control in terms of growth indices and nematode control. In comparison to bitter leaf, the CSK (cashew seed kernel) amendment proved more effective against nematodes. The studies further indicated that both amendments could be used in the nematode control. According to Liman et al. [17], nematode population on tomato seedlings treated with leaf extract from the mahogany plant was considerably lower than on untreated tomato seedlings. They also stated that there are noticeable differences between the extracts' effects on the test organism at different concentrations. The results showed that the severity of galling plant height and root length in tomatoes exposed to various leaf extract concentrations differed significantly.

In the laboratory and the screenhouse, Adenike and Atolani [18] investigated the effectiveness of the *Lawsonia inermis* L. cypress shrub in the management of root-knot nematodes of the genus *Meloidogyne* spp. The quantity of nematode eggs that hatched in the laboratory was dramatically lowered by *L. inermis* aqueous extracts. In comparison to 11.7% of the eggs treated with a 25% crude extract of *L. inermis* (L.), 92% of the control eggs that were not treated hatched. As much as 98.4% of the *L. inermis* (L.) in 100% crude concentration perished within 2 days of the experiment. The amount of nematodes found in the soil and roots of the screenhouse plant was significantly reduced by 15% aqueous crude concentration. Additionally, compared to the untreated control plants, all treated plants galled less.

Olabiyi [19] asserts that a number of plant extracts exhibit nematodal and pathogenic effects on the greenhouse and field pest tomato root-knot nematode. Root-knot nematode eggs were put into tomato seedlings, cv. DT69/257, which were being grown in soil that had been steam-sterilized in the screenhouse at concentrations of 5000, 10,000, 15,000, 20,000, and 25,000. At inocula levels of 15,000, 20,000, and 25,000 eggs of *M. incognita*, the number of leaves, plant height, fruit output, and root galls were all significantly decreased. Field-planted tomato seedlings were treated with aqueous extracts of marigold, nitta, and basil in four different concentrations: 25,000, 500,000, 750,000, and 100,000 ppm (parts per million)/plant. Whole aqueous plant root extracts were used in the trials, and they resulted in a decrease in the number of root-knot nematodes in the soil as well as an increase in plant height, plant leaves, and fruit yield when compared to the control. The aqueous root extracts' ability to effectively control root-knot nematodes was shown by the treated plots' much fewer root galls.

The effectiveness of aqueous ginger extracts against the root-knot nematode *Meloidogyne javanica* was investigated by Amer-Zareen et al. [20]. Higher extract concentrations (100%) in *in vitro* investigations inhibited root-knot egg hatching which led to juvenile death. In a 25% concentration, *Pasteuria* endospore adhesion was stronger. Plant growth was enhanced and disease severity was reduced when *Pratylenchus penetrans* and plant extracts were administered simultaneously. The bacterial antagonist parasitized 72% of the females in the 25% extract plus *P. penetrans* treatment, and each juvenile had nine bacterium spores attached to its nematode cuticle.

According to Dos et al. [21], an ethanolic rhizome extract of *Artemisia vulgaris* suppressed the host plant infectivity, mortality, hatching, and galling of the root-knot nematode *Megadora*. Both of the extract's effects were dose-dependent, with 2.35 mg/ml of the extract reducing egg hatching by 50% and 55.67 mg/ml of the extract raising second-stage juvenile mortality by 50% after 12 h of exposure. An exposed host,

Phaseolus vulgaris and *Becanta trepar*, shows a dose-dependent reduction in nematode infectivity (50% inhibition at 6.28 h). When applied directly to the soil, the extract reduced root-galling on a susceptible host in a dose-dependent manner (50% inhibition at 32.36 mg/ml). While maintained in the dark at 25°C for 15 days, the extract did not lose its activity.

Pavaraj et al. [22] evaluated the efficacy of a bionematicide called goat weed leaf extract against the black gram (*Vigna mungo*)-infested root-knot nematode, *Meloidogyne incognita*. The study also looked at the root gall index to determine the nematode population density. The total protein, lipid, and carbohydrate contents of the experimental plants treated with varied concentrations (2 to 10 ppm) of *Ageratum conyzoides* leaves were also tested after 40 days of treatment (DAT). Because the extract significantly reduced the virulence of the root-knot nematode, it is advisable to utilize it as a bionematicide going forward.

In vitro nematicidal effects of ethanol extracts from the following plant species were reported by Slomp et al. [23]: Apocynaceae species include *Mandevilla velutina* (Mart.) Woodson and *Tithonia diversifolia* (Hemsl.); *Zeyheria montana* Mart. (Bignoniaceae); *Lippia alba* (Mill.) N.E. Brown (Verbenaceae); *Tabernaemontana catharinensis* A. DC.; *Croton antisyphiliticus* Mart. (Euphorbiaceae); and *Serjana erecta* Radlk. (Sapindaceae). The experiment made use of the plant parasitic nematodes such as *Pratylenchus zeae* and *Pratylenchus jaehni*. The findings showed that the studied extracts had considerable nematicidal activity, particularly those that *Eclipta alba* had shown (DL50 (ppm) values = 304.08; 55.32 for *P. zeae* and DL50 (ppm) = 1000; 212.82 for *P. jaehni*, across periods of 12 and 24 h, respectively). *T. catharinensis* had shown (DL50 (ppm) values of 215.26; 60.04 for *P. zeae* and 825.44; 376.60 for *P. jaehni*, across periods of 12 and 24 h, respectively). *Z. montana* had shown (DL50 (ppm) values = 166.43; 34.08 for *P. zeae* and DL50 (ppm) = 1000, 427.34 for *P. jaehni*, across periods of 12 and 24 hours, respectively) and *S. erecta* showed (DL50 (ppm) values = 178.74; 74.12-*P. zeae* and DL50 (ppm) = 689.24; 249.50-*P. jaehni*, across periods of 12 and 24 h, respectively). These results show that the evaluated plant demonstrated significant nematicidal effects, which have significant economic or environmental ramifications and may aid in the extension of agricultural activities around the world.

Under greenhouse and field circumstances, the suppression of the root-knot nematode *M. incognita* was studied by Kamal et al. [24] using plant extracts of eucalyptus (*Eucalyptus camaldulensis*), marigold, garlic, and neem as well as essential oils. After 24–48 h of exposure, an *in vitro* investigation of plant materials demonstrates a nematicidal effect on young worms. The largest percentage of nematode deaths was obtained by the neem extract (65.4%), followed by essential oils (64.4%), marigold extract (60.5%), and garlic and eucalyptus extracts (38.7–39.5%). Neem extracts and essential oil treatments were more effective than other methods at reducing *M. incognita* populations in soil and the root gall index in screenhouse and outdoor settings. Neem and essential oil treatments provided the tomato plant with the most outstanding protection toward root-knot nematode in a field experiment, 44.2 and 32.6%, respectively.

In 2008, Khan et al. assessed the efficiency of ethanolic extracts of *Azadirachta indica*, *Tagetes erecta*, *Withania somnifera*, and *Eucalyptus citriodora* against nematodes associated with papaya (*Carica papaya*). The fresh shoot weight of papaya seedlings grown in pots was significantly increased, according to their research, when plant extracts were used. The fresh root weight was mainly unaltered compared to the control. The amount of incognita juveniles, the root-knot index, and the number of egg masses per root were all considerably reduced by all treatments. The three main nematode species associated with papaya, *M. incognita*, *Helicotylenchus multicinctus*, and *Hoplolaimus indicus*, all significantly lowered their population densities in the field. The most impacted species by *Withania* and *Tagetes* species were *M. incognita, H. multicinctus*, and *H. indicus*. The treatments, in this order: *Withania somnifera* > *carbofuran* > *A. indica* > *T. erecta* > *E. citriodora*, enhanced the papaya yield.

Olabiyi et al. [25] discovered that using aqueous leaf extracts of *Nicotinia tabacum*, *Hyptis suaveolens*, and *C. papaya* at 50 and 100% led to higher growth and yield of tomatoes in comparison to the control using only distilled water. The results also demonstrate that the plant height, number of leaves, root weight, number of fruits, and fruit weight were better at a range of 406 cm, 223, 52, 151, and 534 g, respectively. The control plants were in the following numerical order: 21.3 cm, 11.2, 15.4, 6.5, and 24.5 g. In the control trial, tomato plants were severely galled by *Meloidogyne* species, which led to thick roots. Additionally, the root gall index and the number of soil nematodes were significantly decreased by the leaf extracts.

Meloidogyne incognita egg masses or larvae were subjected to different concentrations of neem leaf (fresh and dry), groundnut leaf, *Borrelia* sp., and garlic bulb, according to Agbenin et al. [26]. The findings demonstrated that compounds from garlic bulbs and neem leaves killed larval stages and prevented egg masses from hatching. Neem and garlic bulb extracts were compared in the screenhouse using weekly applications of 25 ml from each pot at 20% concentrations. In the screenhouse, each pot was infused with 2000 *M. incognita* larvae and 2 kg of pasteurized soil. These extracts significantly reduced tomato root-knot nematode infection indices, as compared to the control. But garlic extract held more promise than neem leaf extract for the *in vivo* control of tomato root-knot infection. Using crushed bark extracts from the *E. camaldulensis, G. arborea*, and *Cassia siamea*, Yusuf [27] demonstrated that they might affect the root-knot nematode's larval egg hatch. He discovered that exposure to these extracts varied in their ability to suppress hatching. As the concentration of the water-soluble bark extracts rises, egg hatch inhibition rises as well. Eighty-nine percent of the larvae hatched much more frequently at the control (distilled water).

In an ultisol solution treated with aqueous leaf extracts of bitter leaf (*Vernonia amygdalina* L.) and mango (*Mangifera indica* L.), the tomato cultivar Roma VF responded to *M. javanica* Treub. infestation, according to a screenhouse experiment conducted by Ogwulumba et al. [28]. There were three different leaf extract concentrations used: 150, 300, and 450 g/l. The results demonstrated that the two leaf extracts considerably altered each of the evaluated parameters. The extracts showed a strong nematotoxic effect on the nematode by reducing the number of galled roots and the galled index as well as increasing the fruit weight at the application rate of 450 g/l.

Susan and Noweer [29] investigated the effects of five aqueous extracts on the root-knot nematode (incognita), including neem seeds (*A. indica*), marigold leaves (*T. erecta*), pyrethrum leaves (*Chrysanthemum cinerariaefolium*), basil leaves (*Ocimum basilicum*), and chinaberry leaves (*Melia azedarach*). All of the materials under investigation affected the survival of the nematode juveniles, depending on the material's properties and concentration. In comparison to Oxamyl 24% L and the untreated control, the majority of the investigated compounds considerably reduced the second-stage juveniles of *M. incognita* in the soil and roots of eggplant (*Solanum melongena* L.) cv. Baladi in field settings. The type and amount of the investigated chemicals had an impact on how much nematodes decreased. The majority of examined substances also resulted in a discernible rise in fruit weight per plant.

According to studies by Saad et al. [30], Khalil [31], Saad et al. [32], and Khan et al. [33], azadirachtin (Achook® 0.15% EC and Nimbecidine® 0.03% EC) showed strong activity against *M. incognita* in tomato plants. Gall, egg masses, and juveniles were reduced by 69.31 and 64.48%, 62.25 and 40.37%.The root-knot nematode, *M. incognita*, on tomato plant was investigated on neem leaves and seeds by the potato cyst nematode (*Globodera rostochiensis*). According to Lynn et al. [34], root-knot nematode (*M. incognita*) on cucumber is significantly suppressed by neem-based formulations and azadirachtin. Neem extracts showed a 38–50% reduction in nematode count [30]. Additionally, investigations on the root-knot worm *M. incognita* showed that the use of dry neem leaves enhanced the weight of fruit per plant of eggplant by 19% [33]. Different dosages of the *Eriobotrya japonica* extract, according to Sultana et al. [35], were successful in preventing *M. incognita* and *Cephalobus litoralis* infections. After 48 h at 1% concentration, the crude extract demonstrated a 90% and ethyl acetate fraction 97% mortality rate against *M. incognita* species, as well as 81 and 50% against *C. litoralis* species.

Izuogu et al. [36] found that root-knot nematodes in maize can be effectively controlled by leaf extracts from *Moringa oleifera* and *Jatropha curcas*. Plants treated with *M. oleifera* perform better in terms of growth parameters than plants treated with *J. curcas* ($P > 0.05$), while untreated control plants experienced the least growth. *A. indica* and *Eucalyptus* spp. seeds were used as aqueous extracts at concentrations of 25, 50, and 100% to control the activity of *M. javanica*. All seed extracts killed *M. javanica* eggs, and second-stage juvenile mortality increased with increasing extract concentration. Liu et al. [37] investigated the impact of a *Dryopteris crassirhizoma* chloroform extract on the ultramicroscopic structures of *M. incognita*. It was shown that the extract significantly damaged the ultrastructure of the worm and clearly damaged the nematode's external and internal structures. According to Jada et al. [38], an ethylacetate extract of *Detarium microcarpum* Guill. and Perr. bark had an impact on the root-knot nematode *M. javanica*. The findings show that the juvenile mortality rate for *M. javanica* was substantially greater at 100% concentration of the ethylacetate extract, with 90 juvenile deaths at 72 h, than at 80% concentration, with 87 juvenile deaths at 72 h. The lowest number of juveniles was found at 0% ethylacetate extract concentration, where one juvenile died after 24 h. Izuogu et al. [39] demonstrated the efficacy of aqueous leaf extracts of *Cassia occidentalis* and *Cymbopogon citratus* at 25, 50, 75, and 100%, while 0% served as the control in the management of the okra nematode. It was determined that the treated plants, especially those that received 50% level and above, performed significantly better ($P = 0.05$) than the control in terms of growth, yield, soil nematode population, root weight, and root gall indices.

Alstonia boonei de Wild and *Bridelia ferruginea* Benth. are two plants whose leaves were examined for nematicidal activity by Fabiyi [40]. With a 75% concentration being the most active, all fractions examined were significantly beneficial in increasing juvenile mortality. With longer exposure times came an increase in mortality. With a percentage mortality of 48.62%, fractions from *Alstonia boonei* were substantially ($P = 0.05$) more toxic to *M. incognita* juveniles than the reference standard carbofuran, which had a mortality rate of 48.89%. Egg hatching was significantly reduced when using the fractions compared to carbofuran's crude extracts. *In vitro* tests were conducted by Tiyagi et al. [41] to determine the effects of water extracts from *Chromolaena odorata, Tithonia diversifolia*, *T. erecta*, and *Occimum gratissimum* leaves at 6.6, 10.0, 13.3, 16.6, and 20% W/V (percent weight/volume) on eggs and second-stage juveniles of incognita. Egg hatch was strongly suppressed by water extracts of *T. erecta* by 90.5% compared to *O. gratissimum*, which generated the lowest egg inhibition of 70.72%. *Tagetes erecta* also caused 100% juvenile mortality within 24 h of exposure, next to *Tagetes diversifolia* (59%), *C. odorata* (50%), and *O. gratissimum* (26.5%) with the minimal concentration.

Aji et al. [42] looked at the effects of several garlic extracts, lemongrass, onion, tridax, and distilled water as control and the results revealed that garlic extract at the crude concentration at 72% h gave the best result (86.68%) followed by its diluted forms. The findings also indicated that nematode mortality increased with exposure time and concentration. The toxic effect of wild sunflower, *Tithonia diversifolia*, was evaluated against the root-knot nematode, *M. incognita*, on eggplant *Solanuum melogena*. *T. diversifolia* aqueous extract was applied at 0, 25, 50 75, and 100% concentrations, while carbofuran was applied at 0, 2500, 5000, 7500, and 1000 ppm. According to the experiment's findings, *T. diversifolia* aqueous extract and carbofuran solution significantly decreased the pace at which nematodes multiply and caused root damage. When compared to the control, this results in higher growth and yield [43].

3. Experimental plants used in the control of plant parasitic nematode

3.1 *Calotropis procera*

Calotropis procera is a member of the Apocynaceae family. The leaves of this untamed shrub, which can reach a height of 1–3 m, are 10–13 cm in width by 17–19 cm in length. It is an established medicinal herb that can be found all across the tropics of Asia and Africa [44]. According to Mother Herbs [45], gigantic swallow wort, often known as milkweed, plays a significant impact in enhancing soil fertility and soil water retention capacity. The root bark is a laxative, febrifuge, anthelmintic, depurative, and expectorant. The root's powder is effective against asthma, bronchitis, and dyspepsia and encourages gastric secretions. The dried whole plant is an excellent tonic, expectorant, depurative, and anthelmintic. In India, the root bark is used to cure leprosy, chronic eczema, and elephantiasis. In addition, paralysis, swellings, arthralgia, and intermittent fever can all be treated with leaves. Flowers are effective in the treatment of tumors, inflammations, catarrh, anorexia, and asthma.

3.2 *Cassia obtusifolia*

According to Brunken et al. [46], the plant *Cassia obtusifolia* L., also known as "sickle pod," is a member of the Fabaceae subfamily Caesalpinioideae. *Senna obtusifolia*, a legume in the *genus* Senna, is another name for it. All of tropical Africa, with the exception of Madagascar, is home to the plant. The species name *obtusifolia* is derived from Latin and means blunt leaf, an accurate description of the shape of the leaves, while the genus name "*Cassia*" came from Greek and signifies an aromatic plant. According to Harvey [47], it is also known as fetid senna or arsenic weed. This plant is grown in home gardens for this purpose in a number of nations, including Senegal, Ghana, Cameroon, and Ethiopia. The young, fragile leaves of this plant are occasionally consumed as vegetables in Africa and worldwide. According to Dirar [48], the plant's green leaves are fermented in Sudan to create the high-protein food item known as "kawal," which is consumed by many as a meat alternative. It is also utilized as a hedge, green manure, medicinal herb, and fuel wood.

3.3 Coffee senna

Coffee senna (*Senna occidentalis* (Linn)) is a member of the Caesalpinioideae subfamily of the Leguminosae family. According to the Greenhouse [49], it is also known as stink wood, smelling pee, Nigerian senna, and Negro coffee. According to Yadava and Satnami [50], this shrub's roots can help prevent ringworm infections. It is used to treat snake bites and as a diuretic. The plant is bitter, laxative, expectorant, anti-inflammatory, antimalarial, and analgesic, according to Mother Herbs [45]. It is used to decrease blood pressure, and bronchitis is treated with a floral infusion and an asthmatic beverage made from the seeds that tastes like coffee.

3.4 *Sesamum indicum*

Sesamum indicum Linn belongs to the Pedaliaceae family. It is a yearly erect herb that is glandular-pubescent and branches from the base. The leaves are alternate or lower appositive and frequently profoundly three lobed [51]. It is cultivated to produce seeds with a high oil content. Most of the world's sesame is grown in tropical, subtropical, and southern temperate countries, as well as in Africa, China, India, and South America. Tropical regions, sandy soil that drains well, and hot, muggy weather are the conditions that the plant grows best in. An excellent source of cooking oil is sesame seeds. Sesame seeds can be used in bread stock, crackers, salad, cooking oil, sesame cakes, wine, and brandy. Young leaves are edible in stews, dried stems are burnable as fuel, and the ash is used to manufacture soap. It is used in confectionery to make sesame seed buns and chips and acts as an insecticide, bactericide, and antifungal synergist for pyrethrum insecticides. Lecithin and lignin give it antioxidant properties and prevent the formation of cholesterol. Sesame oil is used in the pharmaceutical industry to treat nasal mucosa, dryness, impaired vision, vertigo, anxiety, headaches, and sleeplessness. The myristic acid in seeds makes them valuable in the production of medicines that fight cancer. It is also used in the production of biodiesel, a viable alternative to the diesel fuel [52].

3.5 *Waltheria indica*

A flowering plant that belongs to the mallow family, Malvaceae, is called *Waltheria indica* L. This plant's common names include lazy morning velvet leaves and it is pantropical in distribution. It is a short-lived subshrub that can grow up to 2 m tall and has a 2-cm stem diameter. The majority of its habitat is dry, disturbed, and well drained. According to Zongo et al. [53], in Africa, South America, and Hawaii, the herb is often used in traditional medicine, notably to relieve pain, inflammation, asthma, erectile dysfunction, bladder problems, diarrhea, dysentery, conjunctivitis, wounds, and abscesses.

4. Methods of application of the plant extract

4.1 Bare-root dip

Root dipping is the process of coating plant roots with a special solution called root dip, protecting them from pests and diseases, and/or promoting root growth. When used as a root dip to treat *M. incognita*, plant extracts of clove, chili, neem, and onion were found to be beneficial [54]. Chili was found to be the most effective plant extract when treating root-knot nematodes. Leaf and pod extracts of *C. procera* and *Thevetia peruviana* were used as bare-root dip treatments to suppress the phytonematodes *M. incognita* and *Rotylenchulus reniformis* infecting tomato and eggplant. On the treated plants, there were noticeable decreases in the growth of root knots brought on by *M. incognita* and nematode multiplication caused by *R. reniformis*. At various concentrations of leaf extract and dip intensities, *M. incognita* second-stage juvenile larval penetration was likewise impacted. According to Siddiqui and Alam [55], bare-root dipping of tomato and eggplant in leaf extracts of margosa/neem and a related species Persian lilac/bakain significantly reduced root-knot nematode larvae penetration and the ensuing root-galling and population growth of the reniform nematode. The root-dip therapy helped plants grow while masking the harmful nematode effects. Root-galling and the reniform nematode population gradually decreased along with an increase in extract content and dip time. Abbas et al. [56] looked at the efficacy of several application tactics, such as soil drench and root dip in biological remedy (azadirachtin) and synthetic chemicals (Cartap, Virtako), on *M. incognita*. The findings showed that both the treated and untreated root portions of Cartap were more effective than Virtako at reducing egg masses. The quantity of egg masses was dramatically decreased in the treated (45.7, 58.2) and untreated (93.7, 80.7) root portions, respectively, by cure and azadirachtin. When compared to the root-dip approach, the nematode reproduction parameters for soil drench were much lower for all compounds. Jada and Zirafilla [57] used seedling root dips at four different concentrations and four different exposures to determine whether shea butter (*Vitellaria paradoxa*) bark extracts are effective against *M. javanica* infection of tomatoes. The results of the experiment at Loko and Mayo Belwa locations showed that the root-dip method with 60% concentration at 60-min exposure in shea butter bark extracts produced the highest number of flowers/plant, the lowest nematode population in 100 g of soil, and the lowest nematode population in 10 g of roots/plant. They came to the conclusion that root dipping of bark extracts from *V. paradoxa* might control *M. javanica* in the field.

Compared to *T. peruviana*, leaf extracts from *C. procera* caused a relatively stronger suppression of nematode proliferation. Increases in leaf extract content and dip duration resulted in greater reduction in illness incidence [41]. To cure cowpea (*Vigna unguiculata* (L.,) Walp) cultivar pusa komal and okra (*Abelmoschus esculentus* (L.) Moench) infected with *Rotylenchulus reniformis*, bare-root dip treatments of *Murraya koenigii* (L.) Spreng. and *Vitex negundo* L. leaf extract were utilized. *R. reniformis* nematode multiplication on the experimental plants was significantly reduced. Comparatively more nematode proliferation inhibition was induced by Murraya leaf extracts than Vitex. There has been an improvement in plant growth. With an increase in leaf extract content and longer root-dip times, root-dip treatment effectiveness improved in terms of plant weight and decreased disease incidence [58]. On tomato, chili, and brinjal, three natural remedies based on neem oil and pongamia oil (NOPO) were investigated for effectiveness against the root-knot worm *M. incognita*. No. 60 EC (c), No. 60 EC (A), and NOPO 60 EC(c) are the formulations in question. The mixture was evaluated as a root dip for seedlings. The control was outperformed by each of the tree formulations No. 60 EC(c) formulation was shown to be the most effective at reducing the nematode population on tomato, chili, and garden eggs when compared to the other two formulations [59]. When used in three different ways: foliar spray, root dip, and pseudostem injection, the two avermectin chemicals, such as abamectin and emamectin benzoate, have the ability to control plant parasitic nematodes. *Radopholus similis* on banana and *M. incognita* on tomato were the subjects of research. Any of the nematodes tested did not respond well to foliar application of either avermectin or emamectin benzoate to tomato or banana. The effectiveness of banana and tomato root dips in preventing the growth of *M. incognita* in tomatoes as well as *R. similis* in bananas was only average. The common chemical nematicide fenamiphos was not as effective in controlling *M. javanica* and *R. similis* as avermectin injections (1 ml) into banana pseudostems [60]. With the help of corm pairing/hot water root dip, *Tithonia diversifolia* leaf mulch, and a combination of Tithonia and hot water root dip, researchers examined the impact of some cultural practices on the population and control of plant parasitic nematode infestation as well as the impact of these practices on some growth and yield attributes of plantain cv. Agbagba. The untreated plantlets served as the control. According to the findings, *M. incognita, H. multicinctus,* and *R. similis* had the lowest population of plant parasitic nematodes in plots treated with hot water root dip and Tithonia mulch. Tithonia plants' height, leaf count, and sucker production were significantly boosted by a hot water root-dip treatment. According to Nwanguma et al. [61], plantains treated with hot water root dip and Tithonia mulch displayed early flowering and fruiting as well as a noticeable increase in bunch weight.

4.2 Soil drenching

It is the process by which chemicals that are water soluble are injected into the soil and then absorbed systemically by plant components after flooding the roots. It is very useful in combating plant parasitic nematode, fungi, and certain diseases as well as delivering nutrients to the roots [62]. In order to achieve the desired result with soil drenching, the soil should be moist but not saturated [63].

The effectiveness of plant products, which include; neem oil, extract of *Nicotiana tabacum, Veratrum album*, and neem Azal; potato cyst nematode (*G. rostochiensis*) population density and growth, were both detected in response to drenches. All of the formulations that were tested increased plant growth and yield while reducing nematode reproduction. Neem oil combined with extracts of *N. tabacum* and *V. album* at 0.5 and 1.0%, as well as neem Azal at 0.3% (and neem oil at 0.3%), were the most successful in lowering the feared incidence by 78% [64]. Hosta (Host asp) and ferns (*Matteuccia pensylvanica*) infested by *Aphelenchoides fragariae* were successfully controlled by hot water drenching 300 days after treatment (DAT). As compared to the control (25°C), hot water drenching at 70 and 90°C reduced *A. fragariae* in the soil but not in the leaves. Plants bathed in 90°C water had fewer nematode-infected leaves per plant than plants treated with 25 and 70°C water. The host's development parameters were not negatively impacted by hot water treatments. When compared to the control at 150 DAT, boiling water (100°C) sprayed once every month for three consecutive months (April, May, and June) significantly decreased the number of infected leaves and the degree of infection in hosta leaves but not in fern fronds. Similar to how the population of *A. fragariae* was reduced in hosta leaves by 67%, fern fronds by 50%, and soil by 61–98% over a control period of 150 DAT by boiling water (100°C). Despite having no effect on fern growth, a boiling water drench decreased the size and quantity of hosta leaves by 49 and 22%, respectively, in comparison to the untreated control [65, 66]. The soil was exposed to seven different H_2O_2 concentrations (1, 10, 100, 250, 500, 750, and 1000 mM) at different times (24 h before and 24 h after the inoculation time). All hydrogen peroxide (H_2O_2) treatments markedly reduced the worms' egg/g fresh root reproduction rate in comparison to the untreated control. The nematode reproduction was least impacted at 10 mM H_2O_2. Exogenous injection of H_2O_2 may affect nematode reproduction directly as well as indirectly through its ability to cause tomato plants to become resistant to nematode infection [67]. A greenhouse soil bacteria mixture of *Lactobacillus farraginis*, *Bacillus cereus*, and *Bacillus thuringiensis* strains with antinematode activity was tested for its effectiveness on the root-knot nematode. In order to compare two control groups of *M. incognita*, the soil was either doused with sterile distilled water or treated with the broad-spectrum carbamate insecticide carbofuran. The results show that the bacterial mixture can successfully control the roots against nematode.

In order to assess the development of second-stage juveniles (J2) of *M. javanica* on the roots of sensitive tomato cv., Javed et al. [68] utilized neem formulations applied as soil drenching. Tiny-Tim was investigated in a supervised environment. Beginning 7 days after transplant, three different neem preparations—neem cake, aza 5 mg, and aza 10 mg—were soaked in water at the rate of 10 ml per pot, with 4% ethanol-treated plants serving as a control for comparison. The results revealed that the roots of tomato plants were equally permeable to J2, but substantially less so than those of water control plants after being treated in three different neem formulations and ethanol. More effectively than water control, three neem treatments—including aza (5 mg) and aza (10 mg)—halted the development of J2. On the roots of plants treated with aza (10 mg), less J2 were able to develop into immature females than on the roots of ethanol control plants. The plants treated with all three neem formulations and ethanol showed decreased root gall formation when compared to the water control. Aza was found to work best at 10 mg for preventing root nematode infection. *Piliostigma thonningii* leaf aqueous extracts were applied by drenching around eggplants in a field that had been infected with *M. javanica*. The results revealed that the crude extract-treated eggplants had the tallest plants, the most fruits per plant, the highest yield, and the lowest galling indices and final nematode populations [69].

4.3 Soil amendment

According to Davis and Whiting [70], a soil amendment is any material that is supplemented to a soil to improve its physical properties, such as water retention permeability, water infiltration, drainage, aeration, and structure. The goal is to give the roots a better habitat. Three halophytic plant species, including *Tamarindus indica*, *Suaeda fruticosa*, and *Salsola imricata*, were investigated for their ability to control the *M. javanica* infection that affects okra (*Abelmoschus esculentus* (L.) Moench cv. Arka anamika) and eggplant (*Solanum melongena* L. cv. Black beauty). When incorporated into soil at concentrations of 0.3, 0.5, and 1% (W/W), halophytes significantly decreased hatching and, consequently, mortality of second-stage juvenile *in vitro*. They also significantly increased the growth of eggplant and okra, and at higher concentrations (0.5 and 1%) they were able to control root-knot nematode infection [71]. Studies examining the effects of different soil amendments, including grass, ash, and rice husk, on *M. javanica* infestation on Roma tomato (*Lycopersicon esculentum*), revealed notable variations among the amendments in terms of height, days to 50% flowering, leaf number, fruit per plant, and galls per root per plant. The soil should be treated with organic matter (poultry droppings, grass ash, and rice husk (RH) ash) at the rate of 10 to 20 t ha^{-1} for the greatest growth, performance, and management of the tomato root-knot nematode [28]. The nematicidal efficacy of the leaves of four medicinal plants—*Azadiractha indica* A. juss, *Calotropis procera* (Ait) R. Br., *Datura stramonium* L., and *Tagetes erecta* L.—was examined for the purpose of controlling

M. incognita. In comparison to the untreated, all leaf amendments at various dosages considerably enhanced the plant growth characteristics of okra and decreased root-knot infection [72]. Three organic wastes were employed to control the root-knot nematode (*Meloidogyne* spp.) on tomato: sawdust (SD), rice husk (RH), and trash dump (RD). In total, 15, 30, and 45 tons of organic garbage was spread over each acre. As the control, Furadan (3G) was administered at rates of 16, 32, and 64 kg/ha to unaltered plots. According to the findings of the study [73], the RD treatment considerably increased tomato output by 17–100% for RD, 13–84% for SD, and 21–63% for RH.

On the development of potted mung plants and the root-knot worm *M. incognita*, the effects of urea application and soil treatment with nimin (a neem-based product) and neem, castor, and rocket-salad oils as well as soil amendments with urea coatings in varying concentrations of nimin and oil of neem, castor, and rocket salad at 0.02, 0.04, and 0.06 grams per pot improved plant growth and increased chlorophyll content of mung leaves at 1% treated with urea coated with nimin rather than neem oil, castor oil, and rocket salad. This was accomplished by significantly slowing the growth of the root-knot nematode. According to Wani and Yaqub [74], urea-altered soil performs better than soil amended with lesser amounts. According to Parihar et al. [75], an organic amendment has an antagonistic effect on the root-knot nematode (*M. javanica*) that infects bottle gourds. Before inoculating bottle gourds with second-stage juvenile root-knot nematodes, *Datura stramonium* leaves were mixed with the soil. When compared to other plant species or organic additives like *Argemone mexicana*, *Lantana camara*, *Parthenium hyterophrus, W. somnifera*, and others, the stramonium-treated soil (100 g) leavers were found to be most effective at decreasing the reproductive potential of the nematode and increasing chlorophyll content and plant growth.

Numerous plant species' essential oils and extracts have demonstrated promising nematicidal efficacy against nematodes that parasitize plants. For instance, nematode mortality in extracts from *Tagetes* spp. and *M. azedarach* aerial portions was considerable [76, 77]. Other plants with essential oils that contain nematicidal properties include thyme (*Thymus vulgaris*), garlic (*Allium sativum*), and oregano (*Origanum vulgare*) [78, 79]. Garlic aqueous extracts significantly decreased the quantity of *M. incognita* juveniles in a research by Mokbel et al. [80]. These natural nematicides provide prospective substitutions for synthetic chemical nematicides, lowering dangers to human health and environmental contamination.

Green manures and composted plant matter are examples of plant-based supplements that can indirectly enhance soil health and control nematode numbers. According to Borges et al. [81], these additions improve the soil microbial community by encouraging the activity of beneficial species such nematophagous fungus, bacteria, and predatory nematodes that eat plant parasitic nematodes. According to Riga et al. [82], compost application led to a decrease in the number of root-knot nematodes. *Argemone mexicana* L. (Papaveraceae) was studied for its allelochemical and nematicidal potential, and Shaukat et al. [83] found that the polar nature of the toxins was indicated by the fact that an ethanol-soluble extract of the leaf material caused more juvenile mortality of *M. javanica* than either ethyl acetate or hexane extracts did. *A. mexicana* decomposing tissues in soil at 50 g kg1 were extremely harmful and caused 80% plant mortality in tomato plants. Plant development was improved at 10 g kg1, however it was significantly slowed down at 30 g kg1. When *A. mexicana* was allowed to disintegrate in the soil, *M. javanica* population densities in the rhizosphere and in roots, as well as gall formation, were considerably reduced by 10, 30, or 50 g kg1.

Additionally, some green manures alter the nutritional balance of the soil, which has a detrimental effect on nematode populations. For instance, imbalances in the carbon-to-nitrogen (C/N) ratio can be brought on by significant amounts of carbonaceous additives like rice bran, wheat straw, or sawdust. This causes nitrogen to become immobilized, which restricts the amount of nitrogen available to nematodes and hence lowers their number. Desaeger et al. [84] study showed that carbonaceous amendments have a suppressive effect on *M. incognita*. Glucosinolates and thiophenes, which have potent nematicidal effects, have been demonstrated to be released by certain plants, such as mustard (*Brassica* spp.) and marigold (*Tagetes* spp.) [85, 86]. Neem (*A. indica*) and castor (*Ricinus communis*) plants have also demonstrated inhibitory effects on nematode reproduction [87, 88].

In contrast to the control treatment, which saw an increase in nematode population, Kago et al. [89] observed that Brassica tissue treatments at 1908 g and 5292 g and the other treatments inhibited nematode population throughout the third season. The effectiveness of the treatments was evident in seasons two and three due to the sharp decline in nematode numbers. The study also showed that soils supplemented with Brassica tissue contained much more nutrients than the other treatments, with high amounts of calcium, potassium, nitrogen, and phosphorus as well as organic carbon being discovered in soils amended with Brassica tissue in comparison to the other treatments.

In comparison to the control, *M. incognita* race 1 and *M. javanica* populations in tomato plants, root gall populations in olive plants, and final nematode populations were all reduced when composted dry cork was added to the potting mixture at varied ratios [90]. Many soil's physical, chemical, and biological features depend on the amount and quality of organic matter, which is why replenishing organic matter has become an important part of soil health management programs, according to Widmer et al. [91]. They highlighted that the diversity and abundance of free-living and parasitic nematodes are impacted by the addition of organic matter to the soil, rotation crops, cover crops, green manure, and other sources of organic matter. They continued by stating that appropriate organic materials should be used in soil management plans in order to enhance the chemical, physical, and biological characteristics of the soil, as well as to reduce plant parasitic nematodes and soilborne illnesses.

Biofumigation is also a form of soil amendment that involves incorporating plant materials, particularly those from the Brassicaceae family, into the soil to release volatile compounds that exhibit nematicidal activity against various plant parasitic nematodes, including *Meloidogyne* spp. (root-knot nematodes) and *Heterodera* spp. (cyst nematodes). Cruciferous plants, such as mustard (*Brassica juncea*), oilseed radish (*Raphanus sativus*), and Ethiopian mustard (*Brassica carinata*), are rich in glucosinolates, which are converted into isothiocyanates upon tissue disruption. These isothiocyanates can significantly lower nematode populations due to their potent nematotoxic effects [92–94]. Sunhemp considerably reduced root-knot nematode populations in okra crops, according to Wilhelm et al.'s [95] research. *Brassica juncea* (Indian mustard) greatly reduced root-knot nematode populations in okra crops, according to a study by Ntalli et al. [96]. Ghosh et al. [97] found that the usage of mustard cake (*Brassica* spp.) significantly reduced the number of root-knot nematodes in eggplant crops. Additionally, marigold intercropping reduced root-knot nematode populations in tomato crops by 70% [98]. The nematicidal activity of neem (*A. indica*) extracts against *Meloidogyne* spp. was demonstrated by Castagnone-Sereno et al. [99], demonstrating that neem contains azadirachtin, which inhibits the worms' ability to molt and hinders their growth and reproduction. Neem is also believed to contain the compound azadirachtin, which has strong nematocidal activity against a number of nematode species [100].

Buena et al. [101] examined the biofumigant impact of pepper crop residues for regulating *M. incognita* populations in the laboratory by adding 0, 5, 10, and 20 g of pepper crop residue to 500 g of nematode-infested soil. After 20 days at 25°C, pepper crop residues significantly reduced *M. incognita* populations and root gall indices in susceptible tomato cv. Marmande while also increasing soil potassium (K), nitrogen (N), and organic carbon (C). The efficiency of biofumigation using pepper crop residues along with fresh animal manures (with and without plastic cover), methyl bromide, and a control was evaluated using root gall indices on a pepper crop. Subtreatments for grafted and nongrafted susceptible peppers were used in all treatments, except the control. According to this finding, biofumigation using pepper crop wastes is an effective nonchemical method for managing *M. incognita* populations, especially when used in conjunction with plastic cover, nitrogen-rich organic matter, and grafting on resistant pepper. On grafted plants, biofumigation using pepper crop residues and plastic cover, and methyl bromide treatment, with equal effects, root gall indices were lower and yields were higher. Das and Bahera [102] carried out a pot culture study to ascertain the effect of two biofumigants, cabbage and cauliflower leaves, on the population of plant parasitic nematodes infecting okra. In comparison to the untreated control, the experiment reduced the number of root-knot nematodes (40.7%), lance nematodes (40.8–80.1%), spiral nematodes (49.1–79.7%), and stunt nematodes (40.8–81.3%) while improving plant growth parameters, such as shoot length (23.3–54.6%), root length (14.1–46.5%), fresh shoot weight (28.4–81.9%), and fresh root weight (22–38%). With regard to reduction in nematode population and improvement in plant growth metrics, it was discovered that the leaves of both cabbage and cauliflower were comparable.

5. Discussion

Nematode problems are difficult to identify early enough, as the nematode keeps building up in the field over time. Epidemics are frequently the result of this, which calls for prompt and efficient control measures. Nematicides are useful as supplemental therapies or as the last resort when conventional forms of control are ineffective, particularly in high priced crops [103]. The demand for natural pesticides has been driven by the need for a better but faster alternative, and the knowledge that is currently available indicates that plants have great potential as a source of natural pesticides [104].

A significant issue with the usage of synthetic nematicides, in addition to their toxicity to mammals, was their recent finding in groundwater [105]. These terrifying environmental risks are an excellent cause to oppose environmentalists' use of them, which increases enthusiasm for greater ecological ways of control. In this aspect, using organic soil amendment has showed potential. However, there are numerous considerations against using organic soil additives on a wide basis. The two main ones are the slow activities and the bulky materials that must be used in huge numbers [106]. These should not be seen as obstacles to the advancement of this promising strategy; instead, technologies should be created to lessen these pressing issues and enhance the strategy's appeal. For instance, the cheaper price and more widespread availability of amendments compensate for their poorer efficacy as compared to synthesized nematicides. Furthermore, small farmers or horticulturists with low financial means who have access to locally produced wastes will benefit most from the utilization of organic matter [107].

There is a chance that as a result of these treatments, hitherto undiscovered antagonists of plant parasitic nematodes could become more prevalent and reduce nematode populations. Plant extracts and organic soil amendments would have an impact on the intricate soil ecosystem. In their native environment, plant parasitic nematodes are attacked by a variety of soil organisms, including fungi, bacteria, protozoa, other nematodes, and invertebrates [108]. However, nothing is known about how they behave in the field. A number of bacteria and fungi tested for the treatment of specific plant parasitic nematodes showed promise. An additional advantage that calls for more research is the liming effect of plant extracts, such as those from *Acalypha wilkesiana* [109].

6. Conclusion

In conclusion, the idea of biological nematicides is highly appropriate and their development should be promoted because there is a need for the creation of novel compounds that would be environmentally friendly. Additionally, nematologists need to develop a more systematic approach to biocontrol and have a better grasp of the ecology of soil microbes. Additionally, details are required regarding the kind of impact the treatments would have on the composition of the soil community. The development of better control measures might benefit from knowledge of how nematodes and their natural enemies interact in their natural habitat. By better understanding how to employ organic material to manage plant parasitic nematodes, this information will prevent this intriguing idea from being abandoned on dubious reasons. We must allow the concept enough time to develop and produce a compelling answer. In the long term, it might even prove to be a more superior and efficient choice than synthetic nematicides. An appealing solution to the environmental issues caused by the use of synthetic nematicides is the use of organic resources, either as soil amendments or in the form of extracts.

The utilization of organic materials as extracts seems to give a good biological alternative to synthetic nematicides, if properly developed. More field testing needs to be carried out to evaluate the usefulness of these materials in practical situations.

Author details

Mohammed Bukar Aji
Faculty of Agriculture, Department of Crop Protection, Modibbo Adama University, Yola, Adamawa State, Nigeria

*Address all correspondence to: ajimoh@mau.edu.ng

References

[1] Lambert K, Bekal S. The history of nematodes. In: Introduction to Plant-Parasitic Nematodes, Department of Crop Sciences. Urbana, IL: University of Illinois, The American Phytopathology Society (APS); 2002

[2] Luc M, Sikora RA, Bridge J. Plant Parasitic Nematodes in Subtropical and Tropical Agriculture. 2nd ed. Wallingford, Oxfordshire, UK: CABI Publishing CAB International; 2005. p. 871

[3] Esser RP, Tarjan AC, Perry VG. Jesse Roy Christie: The gentleman nematologist. Annual Review of Phytopathology. 1989;**27**(1):41-45

[4] Sato K, Kadota Y, Shirasu K. Plant immune responses to parasitic nematodes. Frontiers in Plant Science. 2019;**10**:1165

[5] Abd-Elgawad M, Askary TH. Impact of phytonematodes on agriculture economy. In: Tarique HA, Paulo RPM, editors. Biocontrol Agents of Phytonematodes. Wallingford, UK: CABI; 2015. pp. 3-49

[6] Savary S, Willocquet L, Pethybridge SJ, et al. The global burden of pathogens and pests on major food crops. Nature Ecology and Evolution. 2019;**3**:430-439. DOI: 10.1038/s41559-018-0793-y

[7] Nicol JM, Turner SJ, Coyne DL, den Nijs L, Hockland S, Tahna MZ. Current nematode threats to world agriculture. In: Jones J, Gheysen G, Fenoll C, editors. Genomics and Molecular Genetics of Plant-Nematode Interactions. Dordrecht: Spinger; 2011. pp. 21-43

[8] Adegbite AA, Agbaje GO. Efficacy of Furadan (Carbofuran) in control of root knot nematode (*Meloidogyne incognita* race 2) in hybrid yam varieties in south western Nigeria. World Journal of Agricultural Science. 2007;**3**(2):256-262

[9] Akhtar M, Malik A. Roles of organic soil amendments and soil organisms in the biological control of plant parasitic nematodes. A Review Bioresource Technology. 2000;**74**:35-47

[10] Wachira PM, Kimenju JW, Okoth SA, Mibey RK. Stimulation of nematode destroying fungi by organic amendments applied in management of plant parasitic nematodes. Asian Journal Plant Science. 2009;**8**(2):153-159

[11] Lu FC. A review of the acceptable daily intakes of pesticides assessed by the World Health Organization. Toxcicol and Pharmacol. 1995;**21**:351-364

[12] Wauchope RD, Butler TM, Hornby AG, Beckers PW, Augustijn-Beckers PW, Burt JP. Pesticides properties database for environmental decision making. In: Review of Environmental Contaminants Toxicology. Vol. 123. 1992. pp. 1-157. PMID: 1732992

[13] Mashela PW, Shimelis HA, Madau FN. Comparison of the efficacy of ground wild cucumber fruits, aldicarb and fenamiphos on suppression of *Meloidogyne incognita* in tomato. Phytopathology. 2008;**156**:264-267

[14] Ononuju CC, Nzenwa PD. Nematicidal effects of some plant extracts on egg hatchability and control of Meloidogyne spp. in cowpea (Vigna unguiculata (L.) Walp). African Journal of Plant Science. 2011;**5**(3):176-182

[15] Umar I, Mohammed BA. Effect of water hyacinth (Eichhornia crassipes

(Mart.) Solms) leaf extract on the juvenile mortality of Meloidogyne incognita. IOSR Journal of Agriculture and Veterinary Sciences (IOSR-JAVS). 2013;**4**:46-48

[16] Umar I, Aji MB. Effect of botanicals in the control of Meloigyne incognita (Kofoid and White) Chitwood on soybean (Glycine max (L) Men). ISOR Journal of Agriculture and Veterinary Sciences (IOSR-JAVS). 2013;**4**(2):43-45

[17] Liman B, Ibrahim M, Ibrahim NT, Rabah AB. Effect of mahogany (Khaya senegalensis L) leaf extract on root-knot nematode of tomatoes (Lycopersicum esculentum L.). Nigerian Journal of Basic and Applied Sciences. 2010;**18**(2):272-276

[18] Adenike FO, Atolani O. Lawsonia inermis in the control of Meloidogyne spp. on Corchorus olitorus. Electronic Journal of Agriculture and Food Chemistry. 2011;**10**(3):2000-2006

[19] Olabiyi TI. Pathogenicity study and nematotoxic properties of some plant extracts on the root knot nematodes pest of tomato, Lycopersicon esculentum (L.) Mill. Plant Pathology Journal. 2008;7(1):45-49

[20] Amer-Zareen ZMJ, Nazir J. Nematicidal activity of ginger and its effects on the efficacy of Pasteuria penetrans for the control of root knot nematodes on tomato. Asian Journal of Plant Sciences. 2003;**2**(11):858-860

[21] Dos SS, Costa DR, Santos MSN, Ryan MF. Effect of Artemisia vulgaris rhizome extracts of hatching, mortality and plant infectivity of Meloidogyne megadora. Journal of Nematology. 2003;**35**(4):437-442

[22] Pavaraj M, Karthikaraj, Rajan MK. Effect of leaf extract of Ageratum conyzoides on the biochemical profile of blackgram, Vigna mungo infected by root knot nematode, Meloidogyne incognita. Journal of Biopesticides. 2010;**3**(1):313-316

[23] Slomp L, Paulo S, Suzelie de Castro F, Sonia Z, Beleboni RO. In vitro nematicidal effects of medicinal plants from Sao Paulo state. Brazil Pharmaceutical Biology. 2009;**47**(3):230-235

[24] Kamal AM, Abo E, Magd El-Morsi A, Abdalgaid MA. Management of tomato root knot nematode Meloidogyne incognita by plant extracts and essential oils. Plant Pathology Journal. 2009;**25**(2):189-192

[25] Olabiyi TI, Adepoju IO, Abolusoro SA, Oyedunmade. Suppression of nematode pests of tomato with aqueous leaf extracts of nitta, tobacco and pawpaw. American-Euroasian Journal of Agronomy. 2011;**4**(2):23-27

[26] Agbenin NO, Emechebe AM, Marley PS, Akpa AD. Evaluation of nematicidal action of some botanicals on Meloidogyne incognita in vivo and in vitro. Journal of Agriculture and Rural Development in the Tropics and Subtropics. 2005;**106**(1):27-39

[27] Yusuf AJ. Nematotoxic effects of three indigenous plants on larvae hatch of root knot nematode (Meloidogyne incognita). Continental Journal Agricultural Science. 2007;**1**:1-7

[28] Ogwulumba SI, Uguokwe KI, Ogbuji RO. Studies on Meloidogyne javanica infestation on Roma tomato (Lycopersicon esculentum mill) under different soil amendments. African Journal of Biotechnology. 2010;**9**(22):3280-3283

[29] Susan AH, Noweer EMA. Management of root knot nematode, Meloidogyne incognita on egg plant with some plant extracts. Egypt Journal of Phytopathology. 2005;**33**(2):65-72

[30] Saad ASA, Massoud MA, Ibrahim HS, Khalil MS. Activity of Nemathorin, natural product and bioproducts againt root-knot nematode on tomatoes. Archives of Phytopathology and Plant Protection. 2012;**45**:955-962

[31] Khalil MSH. Influence of nematicides and certain natural products on the products on the infestation of nematodes attacking tomato plants (Lycopersicon esculentum, Mill) [M.Sc thesis]. Alexandria, Egypt: Alexandria University; 2009

[32] Saad ASA, Massoud MA, Ibrahim HS, Khalil MS. Management study for the root-knot nematodes, Meloidogyne incognita on tomatoes using fosthiazate and Arbuscular mycorrhiza fungus. Journal of Advances in Agricultural Research. 2011;**16**:137-147

[33] Khan MR, Mohiddin FA, Ejaz MN, Khan MM. Management of root-knot disease in egg plant through the application of biocontrol fungi and dry neem leaves. Turkish Journal of Biology. 2012;**36**:161-169

[34] Lynn OM, Song W, Shim J, Kim J, Lee K. Effects of azadirachtin and neem based formulations for the control of sweetpotato whitefly and root-knot nematode. Journal of Korean Society of Applied Biology and Chemistry. 2010;**53**:598-604

[35] Sultana N, Akhtar M, Ferheen S, Kausar R, Khatoon Z, Sultana R, et al. Effect of different concentrations of Eriobotrya japonica extract on control of infection by Meloidogyne incognita and Cephalobus litoralis. Journal of Entomology and Nematology. 2014;**6**(2):27-31

[36] Izuogu NB, Badmos AA, Raji SO. The potency of Moringa Oleifera and Jatropha curcas leaf extracts as control for root-knot nematode in maize (Zea mays). International Journal of Phytofuel and Allied Sciences. 2013;**2**(1):116-124

[37] Liu J, Xie S, Feng J, Cai J. Effects of chloroform extract of Dryopteris crassirhizoma on the ultramicroscopic structures of Meloidogyne incognita. Hindawi the Scientific World Journal. 2013;**2013**:1-7

[38] Jada MY, Reuben KD, Oaya CS. Effect of ethylacetate of Detarium microcarpum Guill and Perr stem bark on root-knot nematode (Meloidogyne javanica). In: Paper Presented at 2nd Biennial Conference of Nigerian Society of Nematologists; 14-16 September 2014; Federal University Abeokuta. (ABST/NISON2014.007). Nigerian Society of Nematologist (NISON); 2014

[39] Izuogu NB, Yakubu LB, Nwabia IW. Efficacy of aqueous leaf extracts of negro coffee (Cassia occidentalis) and lemon grass (Cymbopogon citratus) in the management of nematode pests of okra (Abelmoschus esculentus, L. Moench). In: Nigerian Society of Nematologists 2nd Biennial Conference Programme and Book of Abstracts; 14-16 September 2014; Federal University of Agriculture, Abeokuta. Nigerian Society of Nematologist (NISON); 2014

[40] Fabiyi OA. Nematicidal activation of fractions from Alstonia boonei and Bridelia ferruginea on Meloidogyne incognita in programme and book of abstracts (ABST/NISON 2014/003) of Nigerian Society of Nematologists. In: 2nd Biennial Conference; 14-16 September 2014; Federal University of Agriculture, Abeokuta. Nigerian Society of Nematologist (NISON); 2014

[41] Tiyagi SA, Mahmood I, Rizvi R. Application of some latex-bearing-plants for the management of photo nematode infecting tomato and eggplant. Thai Journal of Agricultural Science. 2009;**42**(4):183-189

[42] Aji MB, Umar I, Maryam YA, Ngwandai P. Nematicidal potential of some plant extract on the control of root-knot nematode. Meloidogyne javanica (Kofoid and White, 1919) Chitwood, 1949. In: Paper Presented at 2nd Biennial Conference of Nigerian Society of Nematologists; 14-16 September 2014; Federal University of Agriculture, Abeokuta. Nigerian Society of Nematologist (NISON); 2014

[43] Abolusoro SA, Adedeji IA. Toxic effect wild sunflower, Tithonia diversifolia on root-knot nematode Meloidogyne incognita. In: Paper Presented at 2nd Biennial Conference; 14-16 September 2014; Federal University Abeokuta. Nigerian Society of Nematologist (NISON); 2014

[44] Ashwani K. Calotropis procera a Plant of Traditional Medicine in India. 2009. Available from: http//www.Science_2.0.com [Accessed: March 24, 2014]

[45] Mother Herbs and Agro Products. Cassia occidentalis. 2014. Available from: http://www.Motherherbs.com [Accessed: April 23, 2014]

[46] Brunken U, Schmidt M, Dressler S, Janssen T, Thiombiano A, Zizka G. West African Plants-A Photo Guide. Frankfurt/Main, Germany: Forschungsinstitut Senckenberg; 2008

[47] Harvey C. Cassia Obtusifolia Huntsville Botanical Garden, West Station, Huntsville 35807. 2012

[48] Dirar H. Kawal, meat substitute from fermented Cassia Obtusifolia leaves. Economic Botany. 1984;**38**(3):342-349

[49] The Greenhouse. The Greenhouse Publication, the Quarterly Magazine of Federal Ministry of Agriculture and Rural Development, Abuja. Vol. 1, No. 1. 2012. pp. 01-86

[50] Yadava RN, Satnami. Chemical constituents from Cassia occidentalis Linn. Indian Journal of Chemistry. 2011;**50**:112-118

[51] Nura S, Adamu AK, Muazu S, Dangora DB. Effect of colchicine-induced mutagenesis on growth and yield of sesame (Sesamum indicum L.). Bayero Journal of Pure and Applied Sciences. 2011;**4**(1):121-125

[52] Anitakumar KR, Pal A, Khanum F, Bawa AS. Nutritional, medicinal and industrial uses of sesame (Sesamum indicum L.) seeds-An overview. Agriculture Conspectus Scientificus. 2010;**75**(4):159-168

[53] Zongo F, Ribuot C, Boumendjel A, Guissou I. Botany, traditional uses, phytochemistry and pharmacology of Watheria Indian L. (Syn Watherian Americans): A review. Journal of Ethnopharmacology. 2013;**148**(1):14-26

[54] Sajid AK, Javed NK, Arm Haq IU, Safdar A. Use of plant extracts as bare dip root treatment for the management of Meloidogyne incognita. Pakistan Journal of Phytopathology. 2011;**23**(1):09-13

[55] Siddiqui MA, Alam MM. Control of root-knot and reniform nematodes by bare-root dip in leaf extracts of margosa and Persian lilac. Zeitschrift für Pflanzenkrankheiten und Pflanzenschutz/Journal of Plant Diseases and Protection. 1988;**95**(2):138-142

[56] Abbas H, Javed N, Kamran M, Khan SA, Abbas H, Haq E, et al. Effect of different application methods on the nematicidal activity of bio and synthetic chemicals against root knot nematode, *Meloidogyne incognita* (Kofoid and White) Chitwood. Pakistan Journal of Zoology. 2020;**52**(6):2027-2242

[57] Jada MY, Zirafilla OA. Effects of tomato (Lycopersicon lycopersicum L.) seedlings

root dip in shea butter (Vitellaria paradoxa) bark extracts on root-knot nematodes (Meloidogyne Javanica) infestation and tomato yield in Adamawa, Nigeria. International Journal of Entomology and Nematology Research. 2021;**5**(1):35-51

[58] Usman A, Mansor. Efficacy of certain botanicals for the management of *Rotylenchus reniformis* infecting okra and cowpea. African Journal of Agricultural Research. 2013;**8**(46):5821-5824

[59] Sivakumar M, Gunsekaran K. Management of root knot nematode in tomato, chilli and brinjal by neem oil formulation. Journal of Biopesticides. 2011;**4**(2):198-200

[60] Jansson RK, Rabatin S. Potential of foliar, dip, and injection application of avermectin for control of plant parasitic nematodes. Journal of Nematology. 1998;**30**(1):65-75

[61] Nwanguma EI, Coyne D, Rotifa I, Adebusoye A. Evaluation of low-input management method for plant parasitic nematodes in plantain cv. Agbagba in south western Nigeria. Nigerian Journal of Horticultural Science. 2008;**13**(1):4-13

[62] Grant LB. What Is Soil Drenching. 2021. Available from: https://www.gardeningknowhow

[63] Fishel MF, Dale GA, Kelin WR. Pesticide Injection and Drenching. University of Florida IFAS Extension Fact Sheet. Gainesville, Florida: UF/IFAS Extension; 2022. pp. 274-2022. DOI: 10.32473/edis-pi274-2022

[64] Trifonova Z, Atanasov A. Control of potato cyst nematode Globodera rostochiensis with some plant extracts and neem products. Bulgarian Journal of Agricultural Science. 2011;**17**(5):623-627

[65] Ganpati BJ, Paarwinders GS. Effectiveness of a hot water drench for the control of foliar nematodes Aphelenchoides fragariae in floriculture. Journal of Nematology. 2004;**36**(1):49-53

[66] Khan A, Sayed M, Shaukat SS, Handoo ZA. Efficacy of four plant extracts on nematodes associated with papaya in Sindh, Pakistan. Nematology Mediterranean. 2008;**36**:93-98

[67] Muwaffag RK. Interaction of root knot nematode (Meloidogyne javanica) and tomato as affected by hydrogen peroxide. Journal of Plant Protection Research. 2008;**48**(2):181-187. DOI: 10.2478/v/0045-008-0021-x

[68] Javed N, Anwar SA, Fyaz S, Khan MM, Asfaq M. Effects of neem formulations applied as soil drenching on the development of root-knot nematode *Meloidogyne javanica* on roots of tomato. Pakistan Journal of Botany. 2008;**40**(2):905-910

[69] Mamman A, Umar I, Malgwi AM, Ojo GT. Efficacy of leaf extract of Piliostigma thonningii for control of root-knot nematode (*Meloidogyne javanica*) on eggplant. Journal of Experimental Agriculture International. 2021;**43**(6):45-52

[70] Davis GJ, Whiting D. Choosing a Soil Amendment. Fort Collins Colarado: Colarado State University Extension; 2023

[71] Abbasi MW, Ahmed N, Zaki MJ, Shahid SS. Soil amendment with halophytes induces physiological changes and reduce root-knot nematode infection in egg plant and okra. Phytopathology Mediterranean. 2010;**49**:352-260

[72] Muhammad AH, Tariq M, Kayani MZ. Efficacy evaluation of Azadiractha indica, Caloropis procera, Datura stramonium and Tagetes erecta against root-knot nematode,

Meloidogyne incognita. Pakistan Journal of Botany. 2011;**43**:197-204

[73] Hassan MA, Chindo PS, Marley PS, Alegbejo MD. Management of root–knot nematode (Metoidogyne spp.) on tomato (Lycopersicon lycopersicum) using organic wastes in Zaria, Nigeria. Plant Protection Science. 2010;**46**(1):34-38

[74] Wani AH, Yaqub B. Control of root-knot nematode, Meloidogyne incognita by urea coated with nimin or other natural oils on mung, Vigna radiata (L.) R. Wilczek. Journal of Biopesticides. 2012;**5**(Supplementary):255-258

[75] Parihar K, Rehman B, Siddiqi MA. Impact of organic additives for sustainable management of root-knot nematode in bottle gourd. Bioscience International. 2012;**1**(4):102-105

[76] Campos DP, Furlanetto C, Stein VC, Noronha CB, Duarte JM, Dallemole-Giaretta R. Nematicidal activity of Melia azedarach extracts on Meloidogyne incognita. Crop Protection. 2022;**152**:105793

[77] Fekry IM, El-Haddad AF, El-Gendi AY. Nematicidal potentiality of some wild Egyptian marigold (Tagetes spp.) essential oils against the root-knot nematode (Meloidogyne incognita) and its role in the management of this nematode in tomato plants. Archives of Phytopathology and Plant Protection. 2021;**54**(7-8):447-463

[78] Aissani N, Mabrouk T, Boulahrouf A, Hambaba L, Moussaoui M. Chemical composition and nematicidal potential of essential oils from Algerian garlic against the root-knot nematode Meloidogyne incognita. Journal of Essential Oil Research. 2020;**32**(4):268-279

[79] Díaz-Ruiz G, Mellado-Mojica E, López-Arredondo D, Campos-García J, López-Bucio J. Nematicidal volatile organic compounds released from Sinorhizobium meliloti strains. Journal of Chemical Ecology. 2019;**45**(11-12):969-979

[80] Mokbel AA, Hashem M, Desoky ESM, Rady MM. Comparative nematicidal efficacy of aqueous extracts of some medicinal plants against root-knot nematode, Meloidogyne incognita. Journal of Applied Pharmaceutical Science. 2018;**8**(10):123-128

[81] Borges LR, Dinardo-Miranda LL, Reis LC, Dombroski JLD. Biological control of plant-parasitic nematodes by soil microorganisms. Biological Control. 2020;**141**:104150. DOI: 10.1016/j.biocontrol.2019.104150

[82] Riga E, Fanelli E, Saia S, Del Re AA, Cramer CS, Cirulli M, et al. Compost Amendment and Soil Tillage Effects. 2015

[83] Shaukat SS, Siddiqui IA, Khan GH, Zaki MJ. Nematicidal and allelopathic potential of *Argemone mexicana*, a tropical weed. Plant and Soil. 2002;**245**:239-247

[84] Desaeger JA, Csinos AS, Overman AJ, Funderburk JE, Noling JW, Gilreath JP. Effect of organic amendments on nematode communities and their impact on control of Meloidogyne incognita on tomato. Journal of Nematology. 2005;**37**(4):449-458

[85] Li S, Ge F, Guo D, Chen Y, Chen L, Li X. Allelopathic potential of glucosinolate compounds in Brassica napus root exudates against Meloidogyne incognita. Journal of Pest Science. 2020;**93**(1):401-410

[86] Saha D, Ansari MS, Al-Hejely W, Sharma M, Kumar J. Glucosinolate from Tagetes patula L. roots and their nematicidal activity against root-knot nematodes (Meloidogyne incognita).

Indian Journal of Agricultural Sciences. 2021;**91**(4):602-605

[87] Hussey RS, Barker KR. A comparison of methods of collecting inocula of Meloidogyne spp., including a new technique. In: Plant Disease Reporter. Vol. 57, No. 12. 1973. pp. 1025-1028

[88] Rani A, Ram M, Choudhary RB, Shukla RN, Meena H. Nematicidal effect of neem cake on root-knot nematode (Meloidogyne incognita) of tomato. Journal of Pharmacognosy and Phytochemistry. 2019;**8**(4):382-384

[89] Kago EK, Kinyua ZM, Okemo PO, Maingi JM. Efficacy of brassica tissue and Chalim™ on control of plant parasitic nematodes. Journal of Biology. 2013;**1**(1):32-38

[90] Nico AI, Jmenez-Diaz RM, Castillo P. Control of root-knot nematodes by composted agro-industrial wastes in potting mixtures. Crop Protection. 2004;**23**:581-587

[91] Widmer TL, Mitkowski NA, Abawi GS. Soil organic matter and management of plant-parasitic nematodes. Journal of Nematology. 2002;**34**(4):289-295

[92] Abid M, Ahmad R, Hafeez FY, Hameed A, Shahid M. Biofumigation: A sustainable approach for pest and weed control. Sustainability. 2018;**10**(5):1433. DOI: 10.3390/su10051433

[93] Nkongolo KK, Adekunle OK, Zhang R, Jiang Y, Zhang Y. Organic amendments and biocontrol agents for the control of plant-parasitic nematodes in vegetable production: A review. Journal of Integrative Agriculture. 2018;**17**(9):2087-2104

[94] Tian Y, Wang X, Qiao K, Han Y, Li X. Effect of different Brassica spp. residues on the activity of root-knot nematode (Meloidogyne incognita) and nematode community in soil. Plant Disease. 2020;**104**(6):1643-1652

[95] Wilhelm M, Wilhelm A, Ali R, Mahajan R, Ntalli N, Tsimidou M, et al. Glucosinolate derived nematicides: A general toxic mechanism? Journal of Plant Diseases and Protection. 2018;**125**(3):251-256. Verlag Eugen Ulmer KG

[96] Ntalli NG, Caboni P, Sheibani MS, Miller AZ. Nematicidal activity of Brassica juncea and Eruca sativa on Meloidogyne spp. and potential for their exploitation as green manures in sustainable crop production. Industrial Crops and Products. 2018;**117**:170-177

[97] Ghosh R, Nagpure A, Sharma S. Management of root-knot nematode (Meloidogyne incognita) in eggplant with botanicals. International Journal of Current Microbiology and Applied Sciences. 2017;**6**(3):2530-2537

[98] Mwamula AO, Okello DK, Nyende AB, Ombor J. The effects of marigold (Tagetes spp.) intercropping on plant-parasitic nematodes and the growth and yield of tomato (Solanum lycopersicum). Journal of Phytopathology and Pest Management. 2019;**6**(2):1-8

[99] Castagnone-Sereno P, Danchin EG, Perfus-Barbeoch L, Abad P, Rosso MN. Diversity and evolution of root-knot nematodes, genus Meloidogyne: New insights from the genomic era. Annual Review of Phytopathology. 2013;**51**:203-220

[100] Nguyen HL, Su NW, Lin KJ, Nguyen VH, Wang SL. Nematicidal potential of neem (Azadirachta indica) extracts and their role in enhancing tomato plant resistance against Meloidogyne incognita. Journal of Nematology. 2020;**52**:1-11. DOI: 10.21307/jofnem-2020-063

[101] Buena AP, García-Álvarez A, Díez-Rojo MA, Ros C, Fernández P, Lacasa A, et al. Use of pepper crop residues for the control of root-knot nematodes. Bioresource Technology. 2007;**98**(15): 2846-2851

[102] Das N, Bahera SK. Management of plant parasitic nematodes in okra by bio-fumigation. Annals of Plant Protection Sciences. 2019;**27**(1):152-155

[103] Gowen SR. Chemical Control of Nematodes: Efficiency and Side-Effects. FAO Plant Production and Protection Paper (FAO). 1997

[104] Zasada IA, Halbrendt JM, Kokalis-Burelle N, LaMondia J, McKenry MV, Noling JW. Managing nematodes without methyl bromide. Annual Review of Phytopathology. 2010;**48**:311-328

[105] Fawole B, Egunjobi OA, Adesiyan SO, Babatola OA, Idowu AA. The biology and control of nematode pests of food crops in Africa. In: Proceedings of the Regional Symposium on Biology and Control of Nematode Pest of Food Crops in Africa. Nigeria: University of Ibadan; 1992. pp. 20-24

[106] Wright DL, Tillman B, Jowers E, Marois J, Ferrell JA, Katsvairo T, et al. Management and Cultural Practices for Peanuts. UF/IFAS Extension publication SS-AGR-74. Gainesville, Florida, USA: University of Florida; 2016

[107] Rotimi MO, Speijer P, Waele DD. On-farm assessment of the influence of oil palm bunch refuse on the growth response of plantain (cv. Agbagba) to parasitic nematodes. Journal of Tropical Forest Resources. 1999;**15**(2):121-129

[108] Misari SM. The biology and control of nematodes pests of food crops in Africa: An essential pre-requisite for food self-sufficiency. In: The Biology and Control of Nematodes Pest Regional Symposium Ibadan Nigeria. Hiswill Inforesources Management Limited; 1992. pp. 26-29. ISBN: 9783187317978978318313

[109] Olaniyi MO. Effect of Acalypha wilkesiana leaf extract, hot and boiling water on plantain growth trend and soil nematode densities. Journal of Biology, Agriculture and Healthcare. 2014;**4**(18):105-115

Chapter 5

Plant Parasitic Nematodes: Insights into the Parasitic Potential, Adaptations and Their Interaction with Other Microorganisms

Himani Sharma and Ashok Kumar Chaubey

Abstract

Plant parasitic nematodes are major pests of the agricultural industry in developing countries. This group is associated with different parts viz., flower, leaves, buds, roots, trunk etc., of approximately all crucial crops of agronomy due to their omnipresent nature. They are categorized as endo, ecto and semi-endoparasite based on the feeding habits. PPNs possess remarkable characteristics especially, parasitic adaptations which help in combating adverse conditions. Furthermore, they interact with other microorganisms (pathogens) forming complex diseases in crops. For effective management of the PPNs through biological control, it is essential to understand their parasitic mechanism, adaptation of J2 stages, feeding mechanism, host-nematode relationship and identification of associated microbiota. This review focuses on the basic biology of PPN, feeding habits, parasitic potential, molecular insights for understanding host-nematode relationship and their management by exploiting the inherent abilities of biocontrol agents.

Keywords: plant parasitic nematodes, parasitism, biocontrol, molecular, pathogen

1. Introduction

Plant parasitic nematodes found in association with plants pose a major threat to the agriculture sector and are responsible for approximately \$80–\$118 billion dollars loss globally [1]. They are obligate parasites of plants which derive their nutrition from them by using their specialized structure viz., stylet. The stylet of PPNs required for feeding and also helps in the establishment of the nematode infection in the host as they used to pierce the plant cell wall for invasion. Furthermore, the stylet also showed connections to the glands present in the pharynx which helps in the production of the molecule's requisite during the infection, invasion and other fundamental processes [2, 3]. These molecules released from the PPN affect the host immune system which ultimately enhances the parasitic association. These nematodes possess the potential to damage any part of the plant, however; roots found to be most susceptible for the PPN colonization especially endoparasites [4, 5]. The parasitic adaptations, feeding behavior and various lifestyle modes of PPN help in the survival in adverse conditions.

1.1 Morphology

Nematodes are the fascinating animals on earth after insects, with many free living and parasitic forms. It's a noteworthy characteristic in case of nematodes that all the conceivable habitats viz., marine, soil and aquatic ecosystems are occupied by this amazing group. Diversity can also be seen in their shape and size as they occur in variable shapes such as melon shape (*Meloidogyne*), reniform (*Rotylenchulusreniformis*), or swollen shape (*Tylenchulussemipenetrans*) etc. and their size ranges from microscopic (*Paratylenchus* 0.18–0.6 mm) to several meters long (*Paralongidorus maximus* ~11.0 mm). The general body structure of nematodes is elongated, cylindrical and tapered at both ends as suggested by their name 'nema' which means thread like [6, 7]. The basic body design consists of an inner alimentary canal which is surrounded by an outer tube of body wall which is commonly referred as 'tube within a tube' structure. The alimentary canal basically comprises gonads and digestive tract. Body wall consists of a variety of muscles viz., ventral and dorsal longitudinal muscles which regulate nematode locomotion as well as these muscles are activated by the dorsal and ventral nerves. Unlined, pressurized, fluid filled cavity which is present between the tubes is known as 'pseudocoelom' formed from the blastula surrounding the gut cavity. Its primary function is to provide the turgor pressure for the entire body and subsequently aids the transfer of oxygen, nutrients and metabolic products. Exoskeleton of nematodes, also known as cuticle, is made up of collagens, glycoproteins, insoluble proteins and lipids.

Plant parasitic nematodes possess an accessory structure known as stylet which is utilized for association with the plants and deriving the nutrients. The shape, size

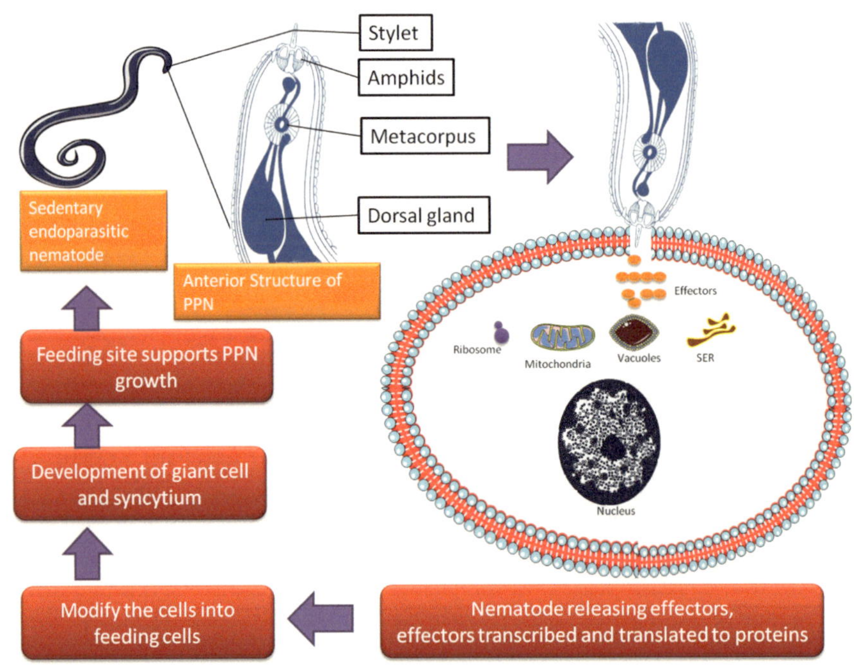

Figure 1.
The anterior end of plant parasitic nematode. The protrusible stylet helps in piercing the plant cell wall. The dorsal gland and other organs produced effectors which modify the surrounding cells for the development of feeding site.

and position of this stylet are of taxonomic importance for the identification of this nemic fauna. The stylet of PPN showed connections with the intestine because of attachment to the pharynx anteriorly (**Figure 1**). Order Tylenchida and Dorylaimida are found to be the pathogen of plants, invertebrates and fungi [8] and as parasites of agricultural crops and forest trees they have great economic importance. All possible habitats or ecological niches have been occupied by Tylenchids. The infection of PPN can be found in all the possible habitats; however, diversity occurred in the root parasites [5]. Over 4000 species of plant parasitic nematodes have been identified but it is very interesting to note that only few genera possess the potential of economic loss viz., *Meloidogyne*, *Helicotylenchus*, *Hoplolaimus*, *Heterodera*, *Globodera*, *Criconemoides*, *Hemicycliophora* etc. [1]. The genus Meloidogyne comprises 105 species, out of which 10 species found to cause potent damage to the agricultural sector [9].

1.2 Life cycle of plant parasitic nematodes

The life cycle of both root knot and cyst nematodes consists of 25–30 days on average. The second juvenile stage (J2) of the PPN finds the host roots for feeding. This stage showed attraction towards the roots and their penetration inside the host cells for satiating the nutrition needs developed by the synctial cells. The synctial comprises the host cells approximately 200 which lost their cell wall and contributed their protoplasts to the large feeding site [10]. After entering into the host cells, the J2 stage molts to J3, J3 molts to J4 and J4 molts to adults. The J2 stage once started feeding becomes larger in size and finally transforms to the males. Among all the larval stages, only the J2 stage showed mobility and others were immobile. Males possess a vermiform body and find their way out from the plants. The exterior environment consists of various chemical cues/pheromones released from the female's body. These cues help males in locating the females for fertilization. Female body comprises eggs produced during fertilization. These eggs were protected in a cyst and all the cysts have the potential to release J2 stage. In case of cyst nematodes, J2 reached to the vascular cells and feeding site developed consists of syncytium cell, however, in root knot nematodes, J2 stage migrated to cortical cells where feeding site developed from the continuous mitosis without following cytokinesis [11]. The process leads to the development of a giant cell which becomes the feeding site for the J2 [12]. The J2 molts thrice and finally becomes an adult. Fertilization is not observed in the root knot, however, males formed during this entire process. Parthenogenesis observed in the root knot nematodes. Eggs are not present inside the female body, instead found outside the body in a protected matrix released from the female body itself. Feeding site either synctial cell or giant cells required repeat stimulus from J2 stage and both nematodes development and life cycle found dependent on these sites (**Figure 2**). This repeated thrust and maintenance of the feeding sites affects the roots which in turn affected the nutrient and water supply to the various parts of the plants, ultimately, affecting the yields.

1.3 Feeding behavior strategies of PPN

PPNs are dependent on the plants for their survival and they are found in close proximity to the plant parts in order to complete their life cycle [6]. They are broadly categorized into above ground feeders and below ground feeders based on their feeding strategies adopted. Further, below ground feeders are classified into ectoparasites, semi-endoparasites and endoparasites.

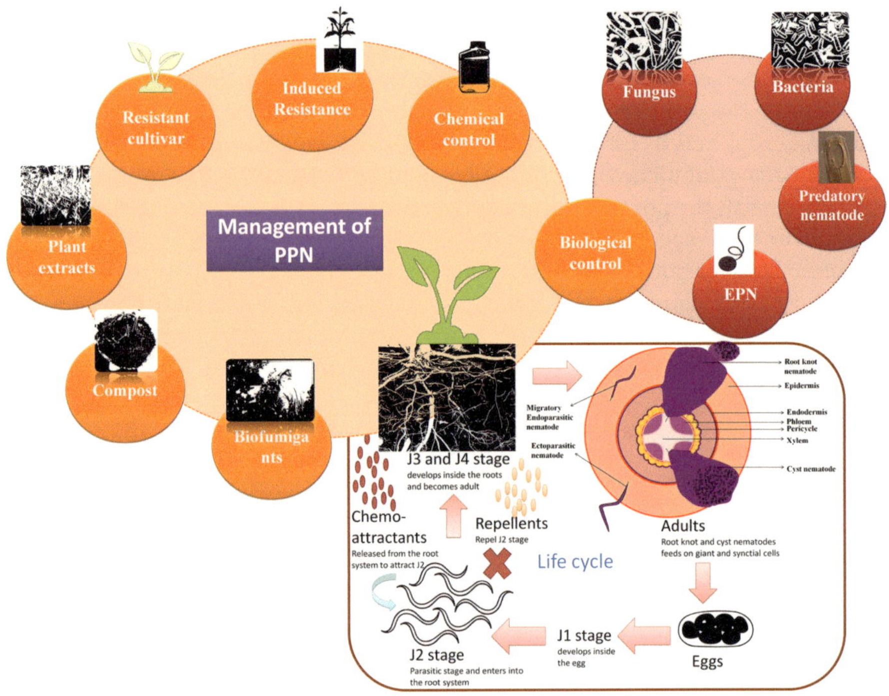

Figure 2.
Different feeding strategies and life cycle of plant parasitic nematode. The role of chemoattrants and repellents on the parasitic J2 stage during the host localization. Management of PPN using different strategies and biological control.

1.3.1 Above ground feeders

1.3.1.1 Stem and bulb nematodes

Ditylenchus dipsac found responsible for causing disease in the plants as it largely affects the stem and bulb of the host hence referred to as stem and bulb nematodes. Hyacinth's brown ring disease caused by this nematode in the plants where wilting, collapsing and yellowish color in the leaves was observed. In the garlic plants, the weight of the bulb decreases and they turn brownish and shriveled. These nematodes affect the storage of these plant bulbs ultimately decreasing the yield [7].

1.3.1.2 Seed gall nematodes

This nematode targets the seeds of the host plant affecting the yield. The disease which affects the seeds of wheat is caused by the Anguinatritici called ear-cockleseed or seed gall, hence commonly referred as seed gall nematode. They are ectoparasites but affect other areas of the plants as soon as it invades the seeds and inflorescence. The diseased plant showed symptoms such as wrinkles, swollen and bulged leaves, whitening, yellowish color of the stems with bending etc. [7].

1.3.1.3 Foliar nematodes

The nematodes belonging to genus *Aphelenchoides spp.* specifically target the leaves of the plants. The symptoms include the yellowish leaves followed by drying and wrinkling, the younger leaves can be spotted with whitish or yellowish patches. Further, it affects the kernels of the plants which become smaller and distortion in their shape is also observed during infection. The grains of the infected plants showed discoloration and became weak due to which cracks were observed [7].

1.3.2 Below ground feeders

1.3.2.1 Ectoparasites

These nematodes found on the surface of the host and derive their nutrition from the root's epidermis. They are further classified into sedentary ectoparasites and migratory ectoparasites. The sedentary nematodes showed specificity in the host and maintain a permanent attachment with the host, for example, Belonolaimus sp., Xiphenema sp., Trichodorus sp. etc. The J2 stage found to be the infective stage in these and only free-living stage. The migratory ectoparasites derive their nutrition from the roots itself but they do not maintain a permanent attachment with the host. Instead, if the roots get disturbed, they will show detachment from the roots and be found in the soil as free nematodes, for example, Criconemella sp., Paratylenchus sp., Hemicycliophora sp. etc. [7].

1.3.2.2 Semi endoparasites

These nematodes were attached with the host anteriorly, however, their posterior body was found free in the soil. They are further classified into sedentary and migratory semi-endoparasites. Insedentary semi-endoparasites, the anterior body showed permanent attachment with the host for example, Tylenchorchynchs, Hoplolaimus etc. TheJ3/J4–adult are the infective stages of sedentary nematodes. The migratory semi-endoparasites does not form any permanent attachment with the host andJ2/J4 are the infective stages, for example, Rotylenchulus sp., Tylenchulus sp. etc. [7].

1.3.2.3 Endoparasites

The endoparasites further categorized into migratory and sedentary. The migratory endoparasites enter into the host through roots and migrate to other areas as they feed on the cells, for exmple, *Pratylenchus* sp, *Radopholus* sp. etc. On the contrary, the sedentary endoparsites also entered through root systems but lost their mobility as they feed on special cells viz., giant cell or syncytial cells, for example, *Meloidogyne* sp., *Heterodera* sp, *Naccobus* sp. The infective stage of sedentary endoparsites is the J2 stage for spreading the infection [7].

2. Adaptation to parasitism

2.1 Parasitic genes and effectors molecule of plant parasitic nematodes

2.1.1 Root-knot nematodes

PPN release effectors molecules to weaken the host immune system. One of the important and harmful root-knot nematodes viz., Meloidogyne releases a number of effectors which helps in their survival by combating the defense system. MiPFN3 gene in Meloidogyne incognita found to be coding for profilin (involved in disassembly of actin) which has the potential to bind the motor actin filament. The effectors molecule binds with the actin filament of the host cell disrupting its normal functioning facilitating the PPN survival [13]. The other effectors are secreted, encoded by gene MilSE5, which disrupts and interrupts important pathways of metabolism [14]. One of the gene viz., Misp12 secreted effectors which directly interfere with the defenses system of the host [15]. The Inflorescence Deficient in Abscission (IDA) like peptide is also encoded by the *M. incognita* geneMilDL1 [16]. The immune system of the host was compromised when effectors, encoded by gene Mh265, released in *M. hapla* [17]. In *M. javanica*, one of the gene MjTTL5 responsible for the production of transthyretin-like protein which helps in suppressing the defense system of the host cell [18]. In *M. graminicola*, effectors and the irrespective genes were identified helping in colonization and combating the defense of the host. The host system was suppressed by the effectors encoded by the genes MgGPP and Mg16820 [19, 20]. The innate immunity is also compromised by the effectors encoded by gene Mg01965 [21]. The apoptosis process in the plant cells were suppressed by the effectors (gene MeTCTP) released by the *M. enterolobii* [22].

2.1.2 Cyst nematodes

Cyst nematodes also released these effectors molecules which enhancing their survival inside the plant's cells and tissues. The gene Hs30D08 in *Heterodera schachtii* codes for the effectors interfering with spliceosome machinery which leads to the changes in those genes responsible for the development of feeding site [23]. Other gene mainly, HsPDI coding for the enzyme disulfide isomerase found to alter the redox potential of plant cell/tissue [24]. Another gene reported from the *H. schachtii* viz., HsTyr encoded tyrosinase enzyme which plays important role in maintenance of homeostasis [25]. The gene HsCLEB found coding for the B-type CLE peptide which plays key role in cell signaling in plants. These peptides had been regulating the vascular cells playing important role in the development of feeding area for the PPN [26]. There are some DNA binding proteins coded by the gene HsGLAND4 which inhibit the transcription of lipid transfer protein as they bind to the promoters observed in *Arabidopsis* [27]. The other gene viz., Hs32E03 coding those molecules which are responsible for the chromatin modifications in order to induce alteration in the ribosomal RNAs of host [28]. There are some effectors which induce alteration in the plant cell wall encoded by the gene HaEXPB2 in *H. avenae* [29]. The immune system of the host found to be compromised especially hypersensitive responses in the presence of effectors coded by HgGLAND18 in *H. glycines* [30]. In *Globodera pallida*, two genes viz., GpSPRY-414-2 and GpGS have been identified for releasing SPRYSEC effector and glutathione synthetase enzyme, respectively [31, 32].

2.1.3 Lesion nematodes

Lesion nematodes, one of migratory parasitic nematodes, releases effector molecules helping in either metabolism or suppression of the host defense system. The gene Ppen12895_c0_seq1 (FAR) encodes for the molecules that play a role in the metabolism of fatty acids in *Pratylenchus penetrans* [33]. Another gene reported from the *P. penetrans* also facilitates survival inside the host [33]. In *P. zeae*, molecules encoded by the gene Vap-1 are found responsible for suppressing the immune system of the host [34]. On the contrary, another gene viz., Sec-2 from the same species played a significant role in combating the defense system of the host [34].

2.1.4 Burrowing nematodes

The burrowing nematode especially, *Radopholus similis* had known to harbor some genes which have facilitated the entry, survival of the nematode and infection to the host. The gene Rs-scp-1 coding the effectors helping the nematode development, facilitating entry into the host and also contribute to the pathogenesis [35]. Another gene viz., Rs-csp responsible for facilitating the juvenile stages entry and also enhances their pathogenicity [36]. Furthermore, the Rs-cb-1 gene is known to code for the molecule playing an essential role during the reproduction and ease invasion of the nematode [37]. Likewise, molecules encoded by the Rs-crt gene facilitate the reproduction and increase pathogenic potential of the nematode [38]. Rs-far-1 gene is found to be a coding effector which contributes during the entry, developmental stages, disrupting the defense system of the host [39].

2.1.5 Potato rot nematodes

In Ditylenchus destructor, two genes DD03093 (VAP-1) and DDC03397 (VAP-2) had been reported for coding the molecules which play an essential role in suppressing the host defense system [40]. The other gene viz., DD03835 (Sec-2) products facilitates the nematodes in overcoming the defense system of the host [40].

2.1.6 Pine wood/wilt nematodes

From the genome of *Bursaphelenchus xylophilus*, six genes have been reported playing a crucial role in parasitism. The first gene, BxSapB1, provides those molecules which enhance the pathogenicity of the nematode and also contribute to the apoptosis of the host cell [41]. One of the genes encodes the 1–3-endoglucanaseenzyme and other encodes expansin-like protein found responsible for degrading the plant cell wall at the time of invasion [42]. To overcome the defense barriers of the host, detoxifying enzymes also released viz., peroxiredoxin, glutathione-S-transferase and cytochrome P450 [43].

2.2 Adaptation of J2 stage during host invasion and colonization

2.2.1 Migratory endoparasites

The juvenile stages of PPNs such as Ditylenchus, Anguina and Pratylenchus require hosts for their survival purposes as they feed on them. Feeding is important for molting which ensures their survival. During extremities of temperature, pH, rain etc. the J2 stage enters into a dormant stage and possesses the ability of surviving without host for a certain period. The quiescent stage in *Radopholus similis* found to be 6 months [44]. The juveniles of Anguina sp withstand the temperature extremities until it finds the host [45]. Some of the PPN juveniles adopt the anhydrobiosis mode which is commonly observed in the entomopathogenic nematodes. The anhydrobiosis process is a gradual release of the moisture from the nematode body to a level where a dried powder form is observed. This process enhances the survival chances of the juveniles for a longer duration without a host. This condition is observed in Ditylenchus sp. The other way of enhancing survival is the arrest of the juvenile stage. This was observed in *Ditylenchus dipsaci* in which the last juvenile stage arrested and showed variations from the normal juveniles. The arrested stage was found to be larger in size with a higher amount of lipid reserves which somehow showed similarity with the *Caenorhabditis elegans* larvae [46]. To understand the similarity between the Ditylenchus and Caenorhabditis, researchers focused on the molecular studies to resolve this mystery. There are some overlapping regions found in this duo which raises the possibility of convergent evolution [47].

2.2.2 Sedentary endoparasites

The migratory stage in these PPNs is the second juvenile stage (J2) which is responsible for spreading the infection in the fields. The hatching of J2 stage from the egg found regulated from various chemical cues released from the specific host. The pre-J2 stage of sedentary nematodes can be quiescent in the environment for a longer period of time until it hatches out from the egg [48]. The signals from the host help in the hatching of the J2 from the egg and they showed similarity with the dauer larvae of free-living nematodes [49]. The J2 stage of cyst nematodes exhibit variations in the cuticle on the basis of lipophilicity as they perceive cues from the host [50]. These J2 stages are either pre- parasitic or parasitic, found to derive their energy from the deposited fat [51, 52]. The larvae of *Heterodera glycines* showed resistance to various harmful toxic substances ensuring its survival [53]. The pre-J2 stage can be dormant for a longer duration; however, the parasitic stage needs to find the suitable host for its survival, development within a short period of time. The parasitic J2 stage hatching should have occurred in the presence of the host. The cues released from the host helps the J2 stages in finding the host. The J2 stages also showed responses towards the chemical signals released from the host parts such as roots and these signals travel larger distances. The other signals possess a shorter range and these signals sensed by the J2 nearer to the host [54]. The J2 stage of *Meloidogyne* hatched out in response to extreme moisture or heat and does not require signals from the plants. This could be the reason behind their broad host range helping them in maintaining their population during unsuitable conditions. Though cyst nematodes J2 stages hatched out in response to environmental conditions, host signals also played an important role. The potato cyst nematode parasitic stage requires specific signals from the host plant for the hatching [55].

2.3 Importance of chemotaxis in host recognition

Previous studies revealed that plant parasitic nematodes especially root knot nematodes (RKN) showed attraction to certain chemicals released from the roots exudates which helps them in finding their specific host [56, 57]. The chemotaxis behavior played an important role in the hatching, survival and development of the plant parasitic nematodes. Though chemicals/compounds released from the host helps in the attraction of J2 stage but some of them work as repellents, some act as stimulants for hatching of the J2 stage and some act as inhibitors [58, 59]. The chemoattractants of RKN found to be released from other parts of the plants also [60]. Likewise, cyst nematodes (CN) also showed chemotactic behavior in response to the compounds secreted from the different parts of the plants [59].

2.3.1 Factors influencing hatching

The hatching in *Heterodera glycines* found stimulated by the compounds Glycinoeclepin A (Pentanor triterpenoid) secreted from the Kidney bean (*Phaseolus vulgaris*) root extracts [54, 61, 62]. 1,10-phenanthroline and tannic acid are synthetic chemicals also known for stimulating hatching in the *H. glycines* [63, 64]. Further, Root exudates of ryegrass (*Lolium multiflorum*) also stimulate the hatching in *H. glycines* [65]. Studies also showed that *H. glycines* cyst wall and egg homogenates and rinsates act as stimulants for the J2 stage hatching from the egg [66]. In *Globodera rostochiensis*, hatching found to be stimulated by the tetranortriterpenoid compound (Solanoeclepin A) and carbohydrate (glucose and fructose) secreted from the *Solanum tuberosum* and *Solanum lycopersicum* root diffusates [67–69]. The compound Asparagusic acid (Dithiolane) secreted from the *Asparagus officinalis* root extract is known to play a role in the inhibition of *H. glycines* and *G. rostochiensis* J2 stage hatching from the egg [70]. The compound allyl isothiocyanate (Isothiocyanate) is a synthetic chemical and formed from the degradation of allyl glucosinolates from *Brassica* plants found to inhibit the *H. glycines* and *H. schachtii* J2 hatching [71]. The root diffusates from *Anthriseussilvestris*, *Arehangeliealitoralis*, *Pimpinella major*, *Pimpinella saxifraga*, *Heracleum sibirieum* acts as hatching inhibitors for the CN mainly *G. rostochiensis* [72]. The synthetic chemical (E)-Chalcone (flavonoid) is also known for inhibiting the *G. rostochiensis* and *G. pallida* J2 hatching [73, 74]. The studies revealed that G. *rostochiensis* J2 hatching can be inhibited from the fermented *Myrothecium verrucaria* and root diffusates from tomato plants with the ABC-G33 and ABC-C6 transporter gene knocked down [75, 76]. The compound Bursehemin and Matairesinol extracted from the leaf of *Bupleurum salicifolium* inhibit the egg hatching in *G. pallida* [70]. The compound allyl isothiocyanate extracted from the Oil of black mustard affected the egg hatching in *G. pallida* [77]. The compounds secreted from the *Bacillus cereus* and *Achromobacter xylosoxidans* also inhibited the hatching of *H. filipjevi* [78].

2.3.2 Chemo attractants

Meloidogyne incognita found attracted to alcohol (Mannitol), organic acids (Vanillic acid, lauric acid), amino acids (Argenine, lysine), phenolics (Tannic acid), phytohormones (6-Dimethylallylaminopurine, salicylic acid, gibberellic acid, Indole-3-acetic acid), organic amines (Cadaverine, 1,3-diaminopropane, putrescine), Calcium chloride [79, 80]. The root diffusates and extracts from marigold, pepper, or soybean seedlings acts as chemoattractants and *H. glycines* J2 stages showed attraction to these [80]. The secondary metabolites of ethylene pathway in *Arabidopsis thaliana* also attracted parasitic stages of *H. schachtii* [81]. The parasitic J2 stage of *G. rostochiensis* is attracted to α-solanine (glycoalkaloids) and linalool [82, 83]. The *G. pallida* J2 stages showed attraction to synthetic compounds viz., indole acetic acid, ethephon, salicylic acid, mannitol, methyl jasmonate, α-solanine and linalool [82, 83].

2.3.3 Repellents

There are some compounds which act as repellents in order to prevent the host from PPN infection. The bulb extracts of *Narcissus tazetta* and metabolites produced during the ethylene pathway in Soybean and *Arabidopsis thaliana* repelled the population of *H. glycines* [84, 85]. The secondary metabolites released during Strigolactone signaling in *Arabidopsis thaliana* are found to repel the population of *H. schachtii* [86]. Further, diffusates from tomato plants root with the ABC-G33 and ABC-C6 transporter gene knocked down in the laboratory conditions revealed the repellent properties against the *G. rostochiensis* and *G. pallida* [76].

3. Management of PPN

3.1 Biocontrol agent

3.1.1 Filamentous fungi

3.1.1.1 Trichoderma

This is considered as one of the important genera in the filamentous fungi for controlling the PPN population. They worked against both root knot and cyst nematodes. They are parasitic to the developmental stages of cyst nematodes. They penetrate the developmental stages by exploiting the enzymes chitinase and protease which break down the extracellular layer. Consequently, the egg number decreases due to fungus parasitism and low level of parasitic stage hatching occurred [87]. In South Africa, Romulus products are prepared from the T. harzianum wettable powder formulation by Dagutat Biolab for controlling the population. In India, there are two commercial products viz., ECOSOM® and commander fungicide prepared from the T. harzianum wettable powder formulation by Agri Life and H.T.C Impex Private Limited, respectively. In Columbia, two commercial products Trichobiol and Trifesol based on the T. harzianum wettable powder formulation were prepared by the Control Biologico Integrado; Mora Jaramillo Arturo Orlando—Biocontrol and Biocultivos Agricultura Sostenible, respectively, as fungal nematicide [88]. The potential of the Trichoderma genus against PPN was presented in **Table 1**.

3.1.1.2 Mycorrhizal fungi

Previous findings on this genus did not report direct protection to the plants against PPN; however, it more efficiently worked on the plants, either by inducing morphological alterations in the roots, providing surplus nutrition or altering the environment for various interactions (**Figure 3**) [97, 98]. Recently, it's been reported that they affect the PPN population and can be used in controlling the PPN infection. The *Rhizophagus intraradices* and *Funneliformis mosseae* found effective in controlling the population of RKN associated with *Solanum lycopersicum* [99].

Fungus	PPN genera	Effects	References
T. longibrachiatum	*Heteroderaavenae*	inhibitory effect on the hatching of cysts	[89]
T. longibrachiatum	*Heteroderaavenae*	Effect on egg, J2 stage and females	[90]
T. harzianum	*M. incognita*	Effect on egg, J2 stage and females	[91]
T. harzianum	*M. incognita*	increased the mortality of the J2	[92]
T. harzianum	*M. incognita*	increased the mortality of the J2	[93]
T. harzianum	*M. incognita*	increased the mortality of the J2	[94]
T. harzianum	*Globodera pallida*	reduced infection and reproduction of the nematode; no effect on the eggs	[95]
T. harzianum	*M. incognita*	decreased eggs hatching and increased the mortality of the J2	[96]

Table 1.
Different species of Trichoderma genus against PPN and their effects.

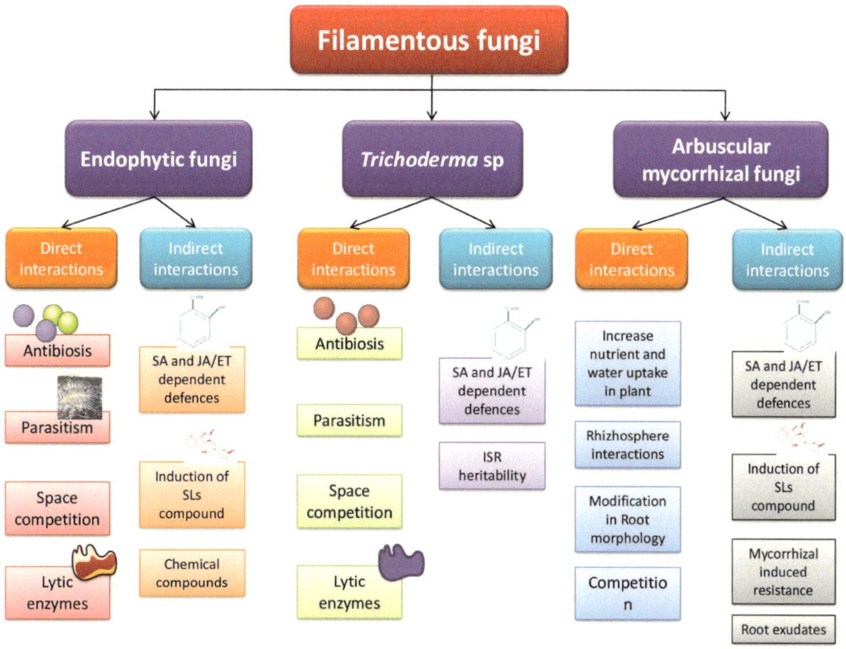

Figure 3.
Direct and indirect interactions of filamentous fungi with root system responsible for inducing induced resistance in plants against the PPN [87]. SA = salicylic acid; JA = jasmonic acid; SLs = strigolactones; ET = ethylene; ISR = induced systemic resistance.

Likewise, *R. intraradices, R. mosseae,* and *R. etunicatum* controlled the population of *Meloidogyne* sp. found in the association with peaches [100]. On the contrary, the increased population of some mycorrhizal fungi enhances the PPN population in the maize field [101]. There are some factors which affect the population of both PPN and fungi and their interaction with each other. The ecological and edaphic factors played a major role in the effects of mycorrhizal fungi against phytoparasitica as the concentration of N and P decreases the biocontrol potential of fungi and increases the population of PPN in the fields [102]. There are contradictions regarding the biocontrol potential of these fungi; however, several studies revealed the biocontrol of these against PPN in different fields [103, 104].

3.1.1.3 Endophytic fungi

The endophytic fungi are also considered for controlling the PPN infection in the field. The studies revealed that these fungi possess the potential of attacking, killing, immobilizing, repelling, interfering with the developmental cells and arouse confusion during host seeking in order to control the PPN population [105]. *Acremonium implicatum* is found regulating the population *M. incognita* present in the rhizosphere of *Solanum lycopersicum* whereas, *F. oxysporum* regulates the *Pratylenchus goodeyi* population in the fields [106]. Both of these fungi worked against the PPN and reduced their population by parasitizing and killing the developmental stages. Furthermore, they produced secondary metabolites against the PPN population which restricts their growth. The *Chaetomium globosum* releases these compounds which affects the growth of Meloidogyne species in the field [107]. *Daldinia cf.*

concentric produced volatile compounds in the field against the *Meloidogyne* species which in turn restrict the growth [108]. The compound of alternariol 9-methyl ether released by the fungi regulates the population of *B. xylophilus* [109]. Though they are promising biocontrol agents in controlling the PPN, some of the entomopathogenic fungus (EPF) found to favor the PPN population. *Pochonia* sp. found to be a potent biocontrol agent in controlling the RKN and CN population. The species of this genus exploited in the formation of commercial products for controlling the PPN infection in the fields. The commercial product KlamiC® (Rothamsted Research and Centro Nacional de Sanidad Agropecuaria) based on granulate formulation from UK & Cuba, PcMR-1 strain (Clamitec, Myco solutions, Lda) based on liquid formulation from Portugal, Xianchongbike (Laboratory for Conservation and Utilization of Bio-resources, Yunnan University) based on liquid formulation from China and IPP-21 product from Italy. *Paecilomyces lilacinus* has been used in commercial products for controlling the PPN population in various countries. The commercial products viz., BIOACT®WP based on water-dispersible powder developed by Bayer Crop Science in USA, PL Gold based on wettable powder formulation by BASF Worldwide, Becker Underwood & PL 251based on water-dispersible granulate by Biological Control Products in South Africa, Stanes Bio Nematon based on liquid or powder Imported from T. Stanes and Company Limited (India) and Gaara Company (Egypt) and BIOCON based on wettable powder developed by Asiatic Technologies Incorporation, Philippines. In India, different companies developed commercial products based on the wettable powder and liquid formulation viz., Shakti Paecil (Shakti Biotech), PAECILO® (Agri Life) and Nematofree (International Panaacea Limited) from India developed for the PPN control. The other formulations from India are Paecilon (Enpro Bio Sciences Private Limited) based on liquid formulation and Gmaxbioguard (GreenmaxAgro Tech) based on talc-based carrier formulation [88].

3.1.1.4 Aspergillus

This genus fungus showed parasitism to the PPN developmental stages and provided protection to the plants [110]. The *Aspergillus niger* control the population of *M. incognita* associated with mung bean, okra and eggplant [111]. Further, *M. javanica* associated with pigeonpea and tomato population was negatively affected in the presence of *A. niger* [112]. The *M. arenaria* population associated with tomatoes was reduced in the presence of *A. niger* [113]. The commercial products of *A. niger* have been developed in India using different formulations. The soluble (liquid) concentrate and suspension concentrate formulation has been used for developing Kalisena products by Cadila Pharmaceutical Limited. The other commercial products viz., Beej Bandhu, Pusa Mrida and Kalasipahi have been developed from the wettable powder and capsule formulation [88].

3.1.2 Bacteria

There are some bacterial species which are found efficacious against the phytoparasitica nematode. The *Pasteuria penetrans* attaches itself to the PPN and their germination leads to the filling of the PPN body with endospores. The *Pseudomonas fluorescens* regulate the population through antibiotics mediating suppression of PPN. The *Bacillus firmus* suppress the population through many ways such as providing protection to the roots, releasing phytohormone, destruction of chemoattractants etc. The *B. thuringiensis* mediates the protection through their proteins which acts against the nematodes. The *B. subtilis* mediates protection through the antibiotics mainly surfactin and iturin. The commercial products of these biocontrol agents were prepared by various companies in different countries. The commercial product based on *P. penetrans* viz., Econem by Nematech Company from Japan and Pasteuria Bioscience from USA. The commercial product based on *Bacillus subtilis* is Stanes Sting (Imported from T. Stanes and Company Limited, India, by Gaara company, Egypt), Quartzo, Nemix C and Presense (FMC Química do Brasil Ltda., Brazil) and Pathway Consortia® (Pathway Holdings, USA) [88].

3.1.3 Entomopathogenic nematodes

Entomopathogenic nematodes (EPNs) gaining attention from the past few years due to their inherent ability of controlling the pest population. However, the characteristics of EPN as controlling agent of pest population also provide safer options from switching harmful insecticides and pesticides [114]. The omnipresent nature of EPN makes them suitable for including in the Integrated Pest Management (IPM) programs. The EPN belongs to two major families that is, Heterorhabditidae and Steinernematidae which includes 19 species from the *Heterorhabditis*, 1 *Neosteinernema* and 126 *Steinernema* species [115, 116]. The parasitic nature of these nematodes observed in the insect's host, however; several present studies suggested that they can also exploit in regulating plant parasitic nematodes population [117]. The detailed list of the PPN controlled by the EPN is provided in **Table 2**.

S.no.	EPN	PPN managed	Crop	Conditions	Effects	References
1.	*S. riobrave*	*M. incognita*	turf grass	*In vivo*	Reduced population	[118]
2.	*Steinernemacar pocapsae*, *S. feltiae*, and *S. riobrave*	*M. incognita*	tomato	Greenhouse trials	Repelling Juveniles	[119]
3.	*S. feltiae*	*M. javanica*	Tomato, Soybeans	Greenhouse trials	reduction in root penetration	[120]
4.	*S. feltiae*	*M. incognita*	tomato	*In vitro*	reduction in egg production and population of J2	[111]
5.	*S. riobrave*	*M. incognita*	tomato	*In vitro*	reduction in egg production and population of J2	[121]
6.	*H. bacteriophora*	*M. incognita*	tomato	*In vitro*	reduction in egg production and population of J2	[121]
7.	*S. riobrave*	*M. partityla*	pecans	Greenhouse trials	reduction in egg masses	[116]
8.	*H. baujardi*	*Meloidogyne mayaguensis*	tomato	*In vitro*	Inhibiting egg hatching and J2 infection	[122]
9.	*S. feltiae*	*Meloidogyne mayaguensis*	tomato	*In vitro*	Inhibiting egg hatching and J2 infection	[122]
10.	*Steinernemararum*	*Nacobbus aberrans*	tomato	Greenhouse trials	reduction in reproduction factor	[123]

S.no.	EPN	PPN managed	Crop	Conditions	Effects	References
11.	*Heterorhabditis bacteriophora*	*Nacobbus aberrans*	tomato	Greenhouse trials	reduction in reproduction factor	[123]
12.	*Heterorhabditis indica*	*M. incognita*	tomato	Greenhouse trials	reduced invasion of *M. incognita*	[124]
13.	*H. bacteriophora*	*M. incognita*	pepper and summer squash	Greenhouse trials	reduction in egg masses	[125]
14.	*S. riobrave*	*M. incognita and M. arenaria*	tomato	direct introduction of IJ/*in vivo*	reduced RKN egg masses	[126]
15.	*S. feltiae*	*M. incognita and M. javanica*	tomato	Greenhouse trials	Decreased infection	[127]
16.	*S. asiaticum, S. glaseri, H. indica and H. bacteriophora*	*M. incognita*	tomato	*In vitro*	reduced number of females	[128]
17.	*S. feltiae*	*M. javanica*	tomato	Greenhouse trials	reduction in egg masses and J2 population density	[117]
18.	*H. bacteriophora*	*M. javanica*	tomato	Greenhouse trials	reduction in egg masses and J2 population density	[117]
19.	*H. bacteriophora*	*M. incognita*	potato	*In vivo*	reduction in reproduction and J2 population density	[129]

Table 2.
The effects of different EPN species against PPN.

3.1.4 Predatory nematodes

The predatory nematodes feed on the plant parasitic nematode and can be exploited in controlling the PPN infection [130]. Most of the predators belong to Mononchida, Dorylaimida, Diplogasterida and Aphelenchida and each of the order possess specialized feeding apparatus [131]. *Clarkuspapillatus* feeds on *Aphelenchoides* sp., *Hemicriconemoides* sp., *Heteroderaschachtii*, *M. hapla*, *T. semipenetrans*, *Subanguinaradicicola* [132]. *Aporcelaimllus nivalis* feeds on *Anguinatritici*, *Aphelenchoides* sp., *Basiria* sp., *Helicotylenchus indicus*, *H. mangiferae*, Heteroderamothi, *H. oryzae*, *Hoplolaimus indicus*, *Longidorus* sp., *M. incognita*, *P. citri*, *Scutellonema* sp., *T. mashhoodi*, *T. semipenetrans*, *Trichodorus* sp., *X. americanum*, *X. insigne* in laboratory conditions [132]. *Aquatidesthornei* found efficacious against *A. tritici*, *H. indicus*, *H. mothi*, *H. oryzae*, *Longidorus* sp., *M. incognita*, *P. citri*, *T. mashhoodi*, *Paratrichodorus* sp., *X. americanum* [132]. Studies on the *Labronemavulvapapillatus* suggested that it feeds on *A. tritici*, *G. rostochiensis*, *M. naasi*, *X. index* in laboratory conditions [132]. The studies on *Mesodorylaimus* sp., *Aporcelaimellus* sp., *Neoactinolaimus* sp., against *R. reniformis*, *R. similis* and *M. incognita* suggested their potential as a biocontrol agent [132]. Despite these experiments, limited studies conducted on the predatory potential of these groups if we compare the publication records from 2000 to 2019 [132]. Their biocontrol potential in greenhouse and in vivo conditions need to be explored in order to use them against the PPN.

3.2 Nematicides

3.2.1 Bionematicides

3.2.1.1 Antagonistic cultivated plants

There are some plants known for the production of compounds which restricts the growth of PPN [133]. These antagonistic/anthelmintic compounds are released in the soil where they regulate the PPN infection and protect the plants. There are so many species that release these anthelmintic compounds but some are found to play an important role in limiting the PPN infection [133]. The genus *Tagetes* releases compounds such as bithienyl and alpha-terthienyl derivatives helping in suppressing the PPN infection. There is another mechanism opted by this genus for controlling the PPN population such as favoring the animal species/plants which are natural predators of nematodes or inhibiting the population, which does not serve for the nutrition requirements to the PPN etc. [134]. The species of this genus found efficacious in controlling the root knot and lesion nematode in tomato plants [133]. *Azadirahta indica* found to be efficient in regulating the nematode population. Further, cruciferous plants have been known for releasing the harmful nematicidal compound. The compound isothiocynates released from these plants when tested against the nematodes found efficacious in population regulation. The usage of this compound as biofumigant against the phytoparasitica nematode was also tested and proved efficient. These compounds are not limited to these plants but other plants also release these volatile molecules acting against the PPN population. The hydrolysed cyanogenic glycoside (cyanide) released from the herbaceous plants found acting against the phytoparasitica nematode [134]. *Crotalaria* species are also exploited for regulating the infection in the fields. The galls produced during root knot nematode infection are found to be decreased in the presence of *Crotalaria longirostrata* by utilizing the harmful compounds released from the microbes [134]. *Zea may* affect the population of lesion nematodes when tested with oats and radish. The nematode population was affected due to the lowering reproductive rate [134]. Likewise, the plant health increases when *Zea mays* cropped with velvet or jack beans and lesion nematode population decreases in response to this [134].

3.2.1.2 Plant-related materials and compounds

There are some compounds isolated from the plants possessing anthelmintic properties that can be used for controlling the PPN infection. These compounds can be prepared from any part of the plant or sometimes whole plant for effective control. The prepared compound can be broadly categorized into acids, oils, alcohol etc. Though, most of these compounds have been prepared from the isolated metabolites of plants but this is not a necessary condition that these will be prepared only from plants. These can be prepared from other microorganisms as they also produce some similar compounds during metabolism [135]. The acetic acid production in the plants occurred during the metabolic pathways but this compound is also produced by the bacterium viz., *Lactobacillus brevis* [136]. This compound possesses the ability to damage the J2stage of root knot nematodes leading to decreased population. Besides this, compounds produced from decomposition of other materials in the soil such as amino acids, butyric acids etc. also cause damage to the phytoparsitica nematode [134]. The compounds released from the *T. viride* and *Secale cereale* affected the PPN infection [134]. Various compounds extracted from the neem which found efficacious against the PPN. Furthermore, the presence of these compounds in the rhizosphere of plants provides protection to the host as the absorption of these compounds in the roots develops resistance against these nematodes. Extensive research has been carried out in this field in order to find the new and effective compounds for management of PPN [134]. The essential oils extracted from the plants were tested against the PPN and differences were found [134]. The essential oils namely carvone, terpinen-4-ol etc. found efficacious against the RKN species in laboratory conditions. The other PPN populations of *R. reniformis*, *Criconemella* and *Hoplolaimus* found resistant against these oils and less mortality observed. The essential oils of *Dysphania ambrosioides* found efficacious in managing these parasitic nematodes [134]. Compounds extracted from the *Paenoiarockii* and *Camellia oleifera* were found efficient in controlling the developmental stages of the RKN [134].

3.3 Host resistance

ISR is an induced systemic resistance in the hosts against the parasites and pathogens causing disease. In this, the host defense system strengthens against the agents' causing diseases through certain compounds [137]. The resistance in the host developed from a particular infection after suffering from the infection earlier and termed as induced resistance [138]. The induced resistance is further classified into two broad categories mainly SAR and ISR [139]. The SAR, that is, Systemic acquired resistance controls the disease-causing agent and decreases the harmful effects of the when compared with other plants. This SAR in the host plants found to be in coordination of the production of salicylic acid during the diseases [140]. The production of salicylic acid mediates the induction of PR-1 gene causing the degradation of pathogen walls. The ISR resistance is found to be regulated by the Jasmonic acid and ethylene pathways. The ISR does not depend upon the salicylic acid and does not show association with the PR genes expression [141]. There are some chemicals identified which provide resistance to the host plants against these parasites [142]. The successful management of lesion and burrowing nematodes found to be based on these chemicals which decrease the PPN infection and enhance the plant growth by providing resistance [143].

4. Conclusions

The PPN infection in the agricultural sectors is one of the major problems in the world. The exploitation of biological control agents especially, fungi, bacteria and EPN could be useful in management of PPN as suggested from research. Further, the use of these agents with other microorganisms or chemical/bionematicide can enhance their efficacy against the PPN. Till now, the use of chemical nematicide against the PPN has some drawbacks which draw the attention towards using the biopesticide. To use EPN agents in the fields it's necessary to develop a successful formulation and extensive studies on their virulence activity against the PPN in *in vivo* condition are required. On the contrary, fungal and bacterial nematicides based commercial products can be easily assessed in the management of PPN.

Conflict of interest

"The authors declare no conflict of interest."

Author details

Himani Sharma* and Ashok Kumar Chaubey
Nematology Laboratory, Department of Zoology, Chaudhary Charan Singh University, Meerut, India

*Address all correspondence to: himanisadhnasharma@gmail.com

References

[1] Bernard GC, Egnin M, Bonsi C. The impact of plant-parasitic nematodes on agriculture and methods of control. Nematology-Concepts, Diagnosis and Control. 2017;**1**:121-151

[2] Jones JT, Haegeman A, Danchin EG, Gaur HS, Helder J, Jones MG, et al. Top 10 plant-parasitic nematodes in molecular plant pathology. Molecular Plant Pathology. 2013;**14**(9):946-961

[3] Mejias J, Truong NM, Abad P, Favery B, Quentin M. Plant proteins and processes targeted by parasitic nematode effectors. Frontiers in Plant Science. 2019;**10**:970

[4] Zinovieva SV. Co-adaptation mechanisms in plant-nematode systems. Parazitologiia. 2014;**48**(2):110-130

[5] Vieira P, Gleason C. Plant-parasitic nematode effectors—Insights into their diversity and new tools for their identification. Current Opinion in Plant Biology. 2019;**50**:37-43

[6] Hartman GL, Rupe JC, Sikora EJ, Domier LL, Davis JA, Steffey KL. Compendium of Soybean Diseases and Pests. St. Paul, MN: American Phytopathological Society; 2015

[7] El-Saadony MT, Abuljadayel DA, Shafi ME, Albaqami NM, Desoky ES, El-Tahan AM, et al. Control of foliar phytoparasitic nematodes through sustainable natural materials: Current progress and challenges. Saudi Journal of Biological Sciences. 2021;**28**(12):7314-7326

[8] Sasser JN, Freckman DW. World perspective on nematology: The role of society. In: Veech JA, Dickson DW, editors. Vistas on Nematology: A Commemoration of the 25th Anniversary of the Society of Nematologists. Hyattsville, Maryland: Society of Nematologists Inc.; 1987. pp. 7-14

[9] Ghaderi R, Karssen G. An updated checklist of Meloidogyne Göldi, 1887 species, with a diagnostic compendium for second-stage juveniles and males. Journal of Crop Protection. 2020;**9**(2):183-193

[10] Jones MG. Host cell responses to endoparasitic nematode attack: Structure and function of giant cells and syncytia. Annals of Applied Biology. 1981;**97**(3):353-372

[11] Von Mende N. Invasion and migration behaviour of sedentary nematodes. In: Fenol C, Grundler FMW, Ohl SA, editors. Cellular and Molecular Aspects of Plant-Nematode Interactions. The Netherlands: Kluwer Academic Publishers; 1997. pp. 51-64

[12] Jagdale S, Rao U, Giri AP. Effectors of root-knot nematodes: An arsenal for successful parasitism. Frontiers in Plant Science. 2021;**12**:800030

[13] Leelarasamee N, Zhang L, Gleason C. The root-knot nematode effector MiPFN3 disrupts plant actin filaments and promotes parasitism. PLoS Pathogens. 2018;**14**(3):e1006947

[14] Shi Q, Mao Z, Zhang X, Zhang X, Wang Y, Ling J, et al. A Meloidogyne incognita effector MiISE5 suppresses programmed cell death to promote parasitism in host plant. Scientific Reports. 2018;**8**(1):7256

[15] Xie J, Li S, Mo C, Wang G, Xiao X, Xiao Y. A novel Meloidogyne incognita effector Misp12 suppresses plant defense

response at latter stages of nematode parasitism. Frontiers in Plant Science. 2016;7:964

[16] Wubben MJ, Gavilano L, Baum TJ, Davis EL. Sequence and spatiotemporal expression analysis of CLE-motif containing genes from the reniform nematode (Rotylenchulusreniformis Linford & Oliveira). Journal of Nematology. 2015;**47**(2):159

[17] Gleason C, Polzin F, Habash SS, Zhang L, Utermark J, Grundler FM, et al. Identification of two *Meloidogyne hapla* genes and an investigation of their roles in the plant-nematode interaction. Molecular Plant-Microbe Interactions. 2017;**30**(2):101-112

[18] Lin B, Zhuo K, Chen S, Hu L, Sun L, Wang X, et al. A novel nematode effector suppresses plant immunity by activating host reactive oxygen species-scavenging system. New Phytologist. 2016;**209**(3):1159-1173

[19] Chen J, Lin B, Huang Q, Hu L, Zhuo K, Liao J. A novel Meloidogyne graminicola effector, MgGPP, is secreted into host cells and undergoes glycosylation in concert with proteolysis to suppress plant defenses and promote parasitism. PLoS Pathogens. 2017;**13**(4):e1006301

[20] Naalden D, Haegeman A, de Almeida-Engler J, Birhane Eshetu F, Bauters L, Gheysen G. The Meloidogyne graminicola effector Mg16820 is secreted in the apoplast and cytoplasm to suppress plant host defense responses. Molecular Plant Pathology. 2018;**19**(11):2416-2430

[21] Zhuo K, Naalden D, Nowak S, Xuan Huy N, Bauters L, Gheysen G. A Meloidogyne graminicola C-type lectin, Mg01965, is secreted into the host apoplast to suppress plant defence and promote parasitism. Molecular Plant Pathology. 2019;**20**(3):346-355

[22] Zhuo K, Chen J, Lin B, Wang J, Sun F, Hu L, et al. A novel Meloidogyne enterolobii effector MeTCTP promotes parasitism by suppressing programmed cell death in host plants. Molecular Plant Pathology. 2017;**18**(1):45-54

[23] Verma A, Lee C, Morriss S, Odu F, Kenning C, Rizzo N, et al. The novel cyst nematode effector protein 30D08 targets host nuclear functions to alter gene expression in feeding sites. New Phytologist. 2018;**219**(2):697-713

[24] Habash SS, Sobczak M, SiddiqueS,VoigtB,ElashryA,GrundlerFM. Identification and characterization of a putative protein disulfide isomerase (HsPDI) as an alleged effector of Heteroderaschachtii. Scientific Reports. 2017;**7**(1):13536

[25] Habash SS, Radakovic ZS, Vankova R, Siddique S, Dobrev P, Gleason C, et al. Heteroderaschachtii Tyrosinase-like protein-a novel nematode effector modulating plant hormone homeostasis. Scientific Reports. 2017;7(1):6874

[26] Vieira P, Eves-Van Den Akker S, Verma R, Wantoch S, Eisenback JD, Kamo K. The Pratylenchus penetrans transcriptome as a source for the development of alternative control strategies: Mining for putative genes involved in parasitism and evaluation of in planta RNAi. PLoS One. 2015;**10**(12):e0144674

[27] Barnes SN, Wram CL, Mitchum MG, Baum TJ. The plant-parasitic cyst nematode effector GLAND4 is a DNA-binding protein. Molecular Plant Pathology. 2018;**19**(10):2263-2276

[28] Vijayapalani P, Hewezi T, Pontvianne F, Baum TJ. An effector from the cyst nematode Heteroderaschachtii derepresses host rRNA genes by altering histone acetylation. The Plant Cell. 2018;**30**(11):2795-2812

[29] Liu J, Peng H, Cui J, Huang W, Kong L, Clarke JL, et al. Molecular characterization of a novel effector expansin-like protein from Heteroderaavenae that induces cell death in Nicotiana benthamiana. Scientific Reports. 2016;**6**(1):35677

[30] Noon JB, Qi M, Sill DN, Muppirala U, Eves-van den Akker S, Maier TR, et al. A plasmodium-like virulence effector of the soybean cyst nematode suppresses plant innate immunity. New Phytologist. 2016;**212**(2):444-460

[31] Mei Y, Wright KM, Haegeman A, Bauters L, Diaz-Granados A, Goverse A, et al. The Globodera pallida SPRYSEC effector Gp SPRY-414-2 that suppresses plant defenses targets a regulatory component of the dynamic microtubule network. Frontiers in Plant Science. 2018;**9**:1019

[32] Lilley CJ, Maqbool A, Wu D, Yusup HB, Jones LM, Birch PR, et al. Effector gene birth in plant parasitic nematodes: Neofunctionalization of a housekeeping glutathione synthetase gene. PLoS Genetics. 2018;**14**(4):e1007310

[33] Vieira P, Maier TR, Eves-van den Akker S, Howe DK, Zasada I, Baum TJ, et al. Identification of candidate effector genes of Pratylenchus penetrans. Molecular Plant Pathology. 2018;**19**(8):1887-1907

[34] Fosu-Nyarko J, Tan JA, Gill R, Agrez VG, Rao U, Jones MG. D e novo analysis of the transcriptome of P ratylenchuszeae to identify transcripts for proteins required for structural integrity, sensation, locomotion and parasitism. Molecular Plant Pathology. 2016;**17**(4):532-552

[35] Huang X, Xu CL, Chen WZ, Chen C, Xie H. Cloning and characterization of the first serine carboxypeptidase from a plant parasitic nematode, Radopholussimilis. Scientific Reports. 2017;**7**(1):4815

[36] Wang K, Li Y, Huang X, Wang DW, Xu CL, Xie H. The cathepsin S cysteine proteinase of the burrowing nematode Radopholussimilis is essential for the reproduction and invasion. Cell & Bioscience. 2016;**6**:1-5

[37] Li Y, Wang K, Xie H, Wang DW, Xu CL, Huang X, et al. Cathepsin B cysteine proteinase is essential for the development and pathogenesis of the plant parasitic nematode Radopholussimilis. International Journal of Biological Sciences. 2015;**11**(9):1073

[38] Li Y, Wang K, Xie H, Wang YT, Wang DW, Xu CL, et al. A nematode calreticulin, Rs-CRT, is a key effector in reproduction and pathogenicity of Radopholussimilis. PLoS One. 2015;**10**(6):e0129351

[39] Zhang C, Xie H, Cheng X, Wang DW, Li Y, Xu CL, et al. Molecular identification and functional characterization of the fatty acid-and retinoid-binding protein gene Rs-far-1 in the burrowing nematode Radopholussimilis (Tylenchida: Pratylenchidae). PLoS One. 2015;**10**(3):e0118414

[40] Peng H, Gao BL, Kong LA, Yu Q, Huang WK, He XF, et al. Exploring the host parasitism of the migratory plant-parasitic nematode Ditylenchus destuctor by expressed sequence tags analysis. PLoS One. 2013;**8**(7):e69579

[41] Hu LJ, Wu XQ, Li HY, Zhao Q, Wang YC, Ye JR. An effector, BxSapB1, induces cell death and contributes to virulence in the pine wood nematode Bursaphelenchusxylophilus. Molecular Plant-Microbe Interactions. 2019;**32**(4):452-463

[42] Shinya R, Morisaka H, Kikuchi T, Takeuchi Y, Ueda M, Futai K. Secretome analysis of the pine wood nematode Bursaphelenchusxylophilus reveals the tangled roots of parasitism and its potential for molecular mimicry. PLoS One. 2013;**8**(6):e67377

[43] Espada M, Silva AC, Eves van den Akker S, Cock PJ, Mota M, Jones JT. Identification and characterization of parasitism genes from the pinewood nematode Bursaphelenchusxylophilus reveals a multilayered detoxification strategy. Molecular Plant Pathology. 2016;**17**(2):286-295

[44] Tarjan AC. Longevity of Radopholussimilis (cobb) in host-free soil. Nematologica. 1961;**6**:170-175

[45] Bird AF, Buttrose MS. Ultrastructural changes in the nematode Anguinatritici associated with an hydrobiosis. Journal of Ultrastructure Research. 1974;**48**(2):177-189

[46] Perry RN, Moens M. Survival of parasitic nematodes outside the host. In: Molecular and Physiological Basis of Nematode Survival. Wallingford UK: CAB International; 2011. pp. 1-27

[47] Van Megen H, Van Den Elsen S, Holterman M, Karssen G, Mooyman P, Bongers T, et al. A phylogenetic tree of nematodes based on about 1200 full-length small subunit ribosomal DNA sequences. Nematology. 2009;**11**:927-950

[48] Hunt DJ, Luc M, Manzanilla-López RH. Plant parasitic nematodes in subtropical and tropical agriculture. In: Luc M, Sikora RA, Bridge J, editors. Identification, Morphology and Biology of Plant Parasitic Nematodes. 2nd ed. Wallingford: CABI publishing; 2005. pp. 11-52

[49] Yang D, Chen C, Liu Q, Jian H. Comparative analysis of pre- and postparasitic transcriptomes and mining pioneer effectors of heteroderaavenae. Cell & Bioscience. 2017;**7**:1-18

[50] Proudfoot L, Kusel JR, Smith HV, Harnett W, Worms MJ, Kennedy MW. Rapid changes in the surface of parasitic nematodes during transition from pre- to post-parasitic forms. Parasitology. 1993;**107**:107-117

[51] McCarter JP, Mitreva MD, Martin J, Dante M, Wylie T, Rao U, et al. Analysis and functional classification of transcripts from the nematode meloidogyne incognita. Genome Biology. 2003;**4**:1-19

[52] Popeijus H, Blok VC, Cardle L, Bakker E, Phillips MS, Helder J, et al. Analysis of genes expressed in second stage juveniles of the potato cyst nematodes globoderarostochiensis and G. pallida using the expressed sequence tag approach. Nematology. 2000;**2**:567-574

[53] Schroeder NE, Mac Guidwin AE. Behavioural quiescence reduces the penetration and toxicity of exogenous compounds in second-stage juveniles of heteroderaglycines. Nematology. 2010;**12**:277-287

[54] Sikder MM, Vestergård M. Impacts of root metabolites on soil nematodes. Frontiers in Plant Science. 2020;**10**:1-18

[55] Perry RN, Clarke AJ. Hatching mechanisms of nematodes. Parasitology. 1981;**83**:435-449

[56] Reynolds AM, Dutta TK, Curtis RH, Powers SJ, Gaur HS, Kerry BR. Chemotaxis can take plant-parasitic nematodes to the source of a chemo-attractant via the shortest possible routes. Journal of the Royal Society Interface. 2011;**8**(57):568-577

[57] Liu W, Jones AL, Gosse HN, Lawrence KS, Park SW. Validation of the

chemotaxis of plant parasitic nematodes toward host root exudates. Journal of Nematology. 2019;**51**(1):1-10

[58] Oota M, Tsai AY, Aoki D, Matsushita Y, Toyoda S, Fukushima K, et al. Identification of naturally occurring polyamines as root-knot nematode attractants. Molecular Plant. 2020;**13**(4):658-665

[59] Ochola J, Coyne D, Cortada L, Haukeland S, Ng'ang'a M, Hassanali A, et al. Cyst nematode bio-communication with plants: Implications for novel management approaches. Pest Management Science. 2021;**77**(3): 1150-1159

[60] Tsai AY, Iwamoto Y, Tsumuraya Y, Oota M, Konishi T, Ito S, et al. Root-knot nematode chemotaxis is positively regulated by l-galactose sidechains of mucilage carbohydrate rhamnogalacturonan-I. Science Advances. 2021;**7**(27):eabh 4182

[61] Masler E, Perry R. In: Perry RN, Moens M, Jones JT, editors. Hatch, Survival and Sensory Perception Cyst Nematodes. Wallingford, UK: C.A.B.I; 2018. pp. 44-73

[62] Bohlmann H. Introductory chapter on the basic biology of cyst nematodes. Advances in Botanical Research. 2015;**73**:33-59

[63] Nonaka S, Katsuyama T, Kondo T, Sasaki Y, Asami T, Yajima S, et al. 1, 10-Phenanthroline and its derivatives are novel hatching stimulants for soybean cyst nematodes. Bioorganic and Medicinal Chemistry Letters. 2016;**26**(21):5240-5243

[64] Renčo M, Sasanelli N, Papajová I, Maistrello L. Nematicidal effect of chestnut tannin solutions on the potato cyst nematode Globoderarostochiensis (Woll.) Barhens. Helminthologia. 2012;**49**:108-114

[65] Riga E, Topp E, Potter J, Welacky T, Anderson T, Tenuta A. The impact of plant residues on the soybean cyst nematode, *Heterodera glycines*. Canadian Journal of Plant Pathology. 2001;**23**(2):168-173

[66] Charlson DV, Tylka GL. Heteroderaglycines cyst components and surface disinfestants affect H. Glycines hatching. Journal of Nematology. 2003;**35**(4):458

[67] Tanino K, Takahashi M, Tomata Y, Tokura H, Uehara T, Narabu T. Total synthesis of Solanoeclepin A. Nature Chemistry. 2011;**3**:484-488

[68] Jones P, Byrne J, Devine K. In vitro studies on the relative availability and mobility in soil of natural hatching factors for the potato cyst nematodes, *Globodera rostochiensis* and *G. pallida*. Nematology. 2001;**3**(1):75-83

[69] Ganapati Reddy P, Chun B-K, Zhang H-R, Rachakonda S, Ross BS, Sofia MJ. Stereoselective synthesis of PSI-352938: A β-D-2′-deoxy-2′-α-fluoro-2′-β-C-methyl-3′,5′-cyclic phosphate nucleotide prodrug for the treatment of HCV. Journal of Organic Chemistry. 2011;**76**:3782-3790

[70] Zhou L, Wang J, Wang K, Xu J, Zhao J, Shan T, et al. Secondary metabolites with antinematodal activity from higher plants. Studies in Natural Products Chemistry. 2012;**37**:67-114

[71] Yu Q, Tsao R, Chiba M, Potter J. Selective nematicidal activity of allyl isothiocyanate. Journal of Food, Agriculture and Environment. 2005;**3**:218-221

[72] Ghisalberti EL. Secondary metabolites with antinematodal activity.

Studies in Natural Products Chemistry. 2002;**26**:425-506

[73] Díaz-Tielas C, Graña E, Reigosa MJ, Sánchez-Moreiras AM. Biological activities and novel applications of chalcones. Planta Daninha. 2016;**34**:607-616

[74] Silva FJ, Campos VP, Oliveira DF, Gomes VA, Barros AF, Din ZU, et al. Chalcone analogues: Synthesis, activity against Meloidogyne incognita, and in silico interaction with cytochrome P 450. Journal of Phytopathology. 2019;**167**(4):197-208

[75] Perry R, Warrior P, Kerry B, Twomey U. Effects of the biological nematicide, DiTera, on hatching of Globoderarostochiensis and G. Pallida. Journal of Nematology. 2000;**2**:355-362

[76] Cox D, Dyer S, Weir R, Cheseto X, Sturrock M, Coyne D, et al. ABC transporters alter plant-microbe-parasite interactions in the rhizosphere. Science Reports. 2019;**9**:526582

[77] Wood C, Kenyon DM, Cooper JM. Allyl isothiocyanate shows promise as a naturally produced suppressant of the potato cyst nematode, Globodera pallida, in biofumigation systems. Nematology. 2017;**19**(4):389-402

[78] Zhang J, Li Y, Yuan H, Sun B, Li H. Biological control of the cereal cyst nematode (Heterodera filipjevi) by Achromobacterxylosoxidans isolate 09X01 and Bacillus cereus isolate 09B18. Biological Control. 2016;**92**:1-6

[79] Fleming TR, Maule AG, Fleming CC. Chemosensory responses of plant parasitic nematodes to selected phytochemicals reveal long-term habituation traits. Journal of Nematology. 2017;**49**(4):462

[80] Wang C, Masler EP, Rogers ST. Responses of Heteroderaglycines and Meloidogyne incognita infective juveniles to root tissues, root exudates, and root extracts from three plant species. Plant Disease. 2018;**102**(9):1733-1740

[81] Kammerhofer N, Radakovic Z, Regis JM, Dobrev P, Vankova R, Grundler FM, et al. Role of stress-related hormones in plant defence during early infection of the cyst nematode Heteroderaschachtii in Arabidopsis. New Phytologist. 2015;**207**(3):778-789

[82] Būda V, Čepulytė-Rakauskienė R. The effects of α-solanine and zinc sulphate on the behaviour of potato cyst nematodes Globoderarostochiensis and G. Pallida. Nematology. 2015;**17**(9):1105-1111

[83] Būda V, Čepulyt Ė-Rakauskien Ė R. The effect of linalool on second-stage juveniles of the potato cyst nematodes Globoderarostochiensis and G. Pallida. Journal of Nematology. 2011;**43**(3-4):149

[84] Hu Y, You J, Li C, Williamson VM, Wang C. Ethylene response pathway modulates attractiveness of plant roots to soybean cyst nematode heteroderaglycines. Scientific Reports. 2017;**7**(1):41282

[85] Hu Y, You J, Li C, Pan F, Wang C. Assessing the effects of water extract of *Narcissus tazetta* bulb on hatching, mortality, chemotaxis and reproduction of the soybean cyst nematode, *Heterodera glycines*. Nematology. 2019;**22**(1):53-62

[86] Escudero Martinez CM, Guarneri N, Overmars H, van Schaik C, Bouwmeester H, Ruyter-Spira C, et al. Distinct roles for strigolactones in cyst nematode parasitism of Arabidopsis roots. European Journal of Plant Pathology. 2019;**154**:129-140

[87] Poveda J, Abril-Urias P, Escobar C. Biological control of plant-parasitic nematodes by filamentous fungi inducers of resistance: Trichoderma, mycorrhizal and endophytic fungi. Frontiers in Microbiology. 2020;**11**:992

[88] Abd-Elgawad MM, Askary TH. Fungal and bacterial nematicides in integrated nematode management strategies. Egyptian Journal of Biological Pest Control. 2018;**28**(1):1-24

[89] Zhang S, Gan Y, Xu B, Xue Y. The parasitic and lethal effects of Trichoderma longibrachiatum against *Heterodera avenae*. Biological Control. 2014;**72**:1-8

[90] Zhang S, Gan Y, Ji W, Xu B, Hou B, Liu J. Mechanisms and characterization of Trichoderma longibrachiatum T6 in suppressing nematodes (*Heterodera avenae*) in wheat. Frontiers in Plant Science. 2017;**8**:1491

[91] Kumar P, Chand R. Bio-efficacy of *Trichoderma harzianum* against root knot nematode *Meloidogyne incognita* on Brinjal. Annals of Plant Protection Sciences. 2015;**23**(2):361-364

[92] Brahma U, Borah A. Management of *Meloidogyne incognita* on pea with bioagents and organic amendment. Indian Journal of Nematology. 2016;**46**(1):58-61

[93] Devi TS, Mahanta B, Borah A. Comparative efficacy of *Glomus fasciculatum*, *Trichoderma harzianum*, *carbofuran* and *carbendazim* in management of *Meloidogyne incognita* and *Rhizoctonia solani* disease complex on brinjal. Indian Journal of Nematology. 2016;**46**(2):161-164

[94] Deori R, Borah A. Efficacy of *Glomus fasciculatum*, *Trichoderma harzianum* for the management of *Meloidogyne incognita* and *Rhizoctonia solani* disease complex in green gram. Indian Journal of Nematology. 2016;**46**(1):61-64

[95] Contina JB, Dandurand LM, Knudsen GR. Use of GFP-tagged *Trichoderma harzianum* as a tool to study the biological control of the potato cyst nematode *Globodera pallida*. Applied Soil Ecology. 2017;**1**(115):31-37

[96] Khan MR, Ahmad I, Ahamad F. Effect of pure culture and culture filtrates of *Trichoderma* species on root-knot nematode, *Meloidogyne incognita* infesting tomato. Indian Phytopathology. 2018;**71**:265-274

[97] Schouteden N, De Waele D, Panis B, Vos CM. *Arbuscular mycorrhizal* fungi for the biocontrol of plant-parasitic nematodes: A review of the mechanisms involved. Frontiers in Microbiology. 2015;**6**:1280

[98] Wani KA, Manzoor J, Shuab R, Lone R. Arbuscular mycorrhizal fungi as biocontrol agents for parasitic nematodes in plants. In: Mycorrhiza-nutrient uptake, biocontrol, ecorestoration. Cham: Springer; 2017. pp. 195-210

[99] Marro N, Caccia M, Doucet ME, Cabello M, Becerra A, Lax P. Mycorrhizas reduce tomato root penetration by false root-knot nematode *Nacobbus aberrans*. Applied Soil Ecology. 2018;**124**:262-265

[100] Calvet C, Pinochet J, Hernández-Dorrego A, Estaún V, Camprubí A. Field microplot performance of the peach-almond hybrid GF-677 after inoculation with *Arbuscular mycorrhizal* fungi in a replant soil infested with root-knot nematodes. Mycorrhiza. 2001;**10**:295-300

[101] Brito OD, Hernandes I, Ferreira JC, Cardoso MR, Alberton O, Dias-Arieira CR. Association between *Arbuscular mycorrhizal* fungi and

Pratylenchus brachyurus in maize crop. Chilean Journal of Agricultural Research. 2018;**78**(4):521-527

[102] Ferreira BS, Santana MV, Macedo RS, Silva JO, Carneiro MA, Rocha MR. Co-occurrence patterns between plant-parasitic nematodes and arbuscular mycorrhizal fungi are driven by environmental factors. Agriculture, Ecosystems and Environment. 2018;**265**:54-61

[103] Emery SM, Reid ML, Bell-Dereske L, Gross KL. Soil mycorrhizal and nematode diversity vary in response to bioenergy crop identity and fertilization. GCB Bioenergy. 2017;**9**(11):1644-1656

[104] Alvarado-Herrejón M, Larsen J, Gavito ME, Jaramillo-López PF, Vestberg M, Martínez-Trujillo M, et al. Relation between arbuscular mycorrhizal fungi, root-lesion nematodes and soil characteristics in maize agroecosystems. Applied Soil Ecology. 2019;**135**:1-8

[105] Schouten A. Mechanisms involved in nematode control by endophytic fungi. Annual Review of Phytopathology. 2016;**54**:121-142

[106] Mwaura P, Dubois T, Losenge T, Coyne D, Kahangi E. Effect of endophytic *Fusarium oxysporum* on paralysis and mortality of *Pratylenchus goodeyi*. African Journal of Biotechnology. 2010;**9**(8):1130-1134

[107] Khan B, Yan W, Wei S, Wang Z, Zhao S, Cao L, et al. Nematicidal metabolites from endophytic fungus Chaetomium globosum YSC5. FEMS Microbiology Letters. 2019;**366**(14): fnz 169

[108] Liarzi O, Bucki P, Braun Miyara S, Ezra D. Bioactive volatiles from an endophytic Daldinia cf. concentrica isolate affect the viability of the plant parasitic nematode *Meloidogyne javanica*. PLoS One. 2016;**11**(12):e0168437

[109] Lou J, Yu R, Wang X, Mao Z, Fu L, Liu Y, et al. Alternariol 9-methyl ether from the endophytic fungus Alternaria sp. Samif 01 and its bioactivities. Brazilian Journal of Microbiology. 2016;**47**:96-101

[110] Askary TH. Nematophagous fungi as biocontrol agents of phytonematodes. In: Biocontrol Agents of Phytonematodes. Wallingford, UK: CABI; 2015

[111] Bhat MY, Wani AH. Bio-activity of fungal culture filtrates against root-knot nematode egg hatch and juvenile motility and their effects on growth of mung bean (Vigna radiata L. Wilczek) infected with the root-knot nematode, *Meloidogyne incognita*. Archives of Phytopathology and Plant Protection. 2012;**45**(9):1059-1069

[112] Zareen A, Zaki MJ, Khan NJ. Effect of fungal filtrates of aspergillus species on development of root-knot nematodes and growth of tomato (Lycopersicon esculentum mill). Pakistan Journal of Biological Sciences. 2001;**4**(8):995-999

[113] Mokbel AA, Obad IM, Ibrahim IK. The role of antagonistic metabolites in controlling root-knot nematode, *Meloidogyne arenaria* on tomato. Alexandria Journal of Agricultural Research. 2009;**54**(1):199-205

[114] Khan RR, Ali RA, Ali A, Arshad M, Majeed S, Ahmed S, et al. Compatibility of entomopathogenic nematodes (Nematoda: Rhabditida) and the biocide, spinosad for mitigation of the armyworm, Spodoptera litura (F.) (Lepidoptera: Noctuidae). Egyptian Journal of Biological Pest Control. 2018;**28**:1-6

[115] Sharma H, Rana A, Bhat AH, Chaubey AK. Entomopathogenic nematodes: Their characterization, bio-control properties and new perspectives. In: Nematodes-Recent Advances, Management and New Perspectives. London: Books on Demand; 2021. p. 187

[116] Malan AP, Knoetze R, Kapp C, Tiedt L, Steinernemabakwenae n. sp. (Rhabditida: Steinernematidae), a new entomopathogenic nematode from South Africa. Nematology. 2023;**25**:275-293

[117] El Aimani A, Houari A, Laasli SE, Mentag R, Iraqi D, Diria G, et al. Antagonistic potential of Moroccan entomopathogenic nematodes against root-knot nematodes, *Meloidogyne javanica* on tomato under greenhouse conditions. Scientific Reports. 2022;**12**(1):2915

[118] Grewal PS, Martin WR, Miller RW, Lewis EE. Suppression of plant-parasitic nematode populations in turfgrass by application of entomopathogenic nematodes. Biocontrol Science and Technology. 1997;**7**:393-399

[119] Grewal PS, Lewis EE, Venkatachari S. Allelopathy: A possible mechanism of suppression of plant-parasitic nematodes by entomopathogenic nematodes. Nematology. 1999;**1**(7):735-743

[120] Fallon DJ, Kaya HK, Gaugler R, Sipes BS. Effects of etomopathiogenic nematodes on *Meloidogyne javanica* on tomatoes and soybeans. Journal of Nematology. 2002;**34**(3):239

[121] Pérez EE, Lewis EE. Use of entomopathogenic nematodes to suppress *Meloidogyne incognita* on greenhouse tomatoes. Journal of Nematology. 2002;**34**(2):171

[122] Molina JP, Dolinski C, Souza RM, Lewis EE. Effect of entomopathogenic nematodes (Rhabditida: Steinernematidae and Heterorhabditidae) on *Meloidogyne mayaguensis* Rammah and Hirschmann (Tylenchida: Meloidoginidae) infection in tomato plants. Journal of Nematology. 2007;**39**(4):338

[123] Caccia M, Lax P, Doucet ME. Effect of entomopathogenic nematodes on the plant-parasitic nematode *Nacobbus aberrans*. Biology and Fertility of Soils. 2013;**49**:105-109

[124] Javed N, Khan SA, Atiq M, Kamran M. Effect of *Steinernema glaseri* and *Heterorhabditis indica* on the plant vigour and root knot nematodes in tomato roots at different densities and time of applications. Pakistan Journal of Zoology. 2012;**44**(4):1165-1170

[125] del Valle EE, Lax P, Rondán Dueñas J, Doucet ME. Effects of insect cadavers infected by *Heterorhabditis bacteriophora* and *Steinernema diaprepesi* on *Meloidogyne incognita* parasitism in pepper and summer squash plants. Agricultural Science and Research. 2013;**40**:109-118

[126] Kepenekci I, Hazir S, Lewis EE. Evaluation of entomopathogenic nematodes and the supernatants of the *in vitro* culture medium of their mutualistic bacteria for the control of the root-knot nematodes *Meloidogyne incognita* and *M. arenaria*. Pest Management Science. 2016;**72**(2):327-334

[127] Kepenekci I, Hazir S, Oksal E, Lewis EE. Application methods of *Steinernema feltiae*, *Xenorhabdusb ovienii* and *Purpureocillium lilacinum* to control root-knot nematodes in greenhouse tomato systems. Crop Protection. 2018;**108**:31-38

[128] Khan SA, Javed N, Kamran M, Abbas H, Safdar A, ul Haq I. Management of *Meloidogyne incognita* race 1 through the use of entomopathogenic nematodes in tomato. Pakistan Journal of Zoology. 2016;**48**(3):763

[129] Osman HA, Ameen HH, Hammam MM, El-Sayed GM, Elkelany US, Abd-Elgawad MM. Antagonistic potential of an Egyptian entomopathogenic nematode, compost and two native endophytic bacteria isolates against the root-knot nematode (*Meloidogyne incognita*) infecting potato under field conditions. Egyptian Journal of Biological Pest Control. 2022;**32**(1):137

[130] Esser E. Biological control of nematodes by nematodes I. Dorylaims (Nematoda: Dorylaimida). Nematology Circular. 1987;**144**:4

[131] Pervez R, Bilgrami AL, Yoshiga T, Kondo E. Feeding, attraction and aggregation behaviour of *Mesodorylaimus bastiani* and *Aquatides thornei* at the feeding site using *Hirschmanniella oryzae* as prey. International Journal of Nematology. 2000;**10**(2):207-214

[132] Kanwar RS, Patil JA, Yadav S. Prospects of using predatory nematodes in biological control for plant parasitic nematodes—A review. Biological Control. 2021;**160**:104668

[133] Grubišić D, Uroić G, Ivošević A, Grdiša M. Nematode control by the use of antagonistic plants. Agriculturae Conspectus Scientificus. 2018;**83**(4):269-275

[134] Abd-Elgawad MM. Optimizing safe approaches to manage plant-parasitic nematodes. Plants. 2021;**10**(9):1911

[135] Ntalli N, Ratajczak M, Oplos C, Menkissoglu-Spiroudi U, Adamski Z. Acetic acid, 2-undecanone, and (E)-2-decenal ultrastructural malformations on *Meloidogyne incognita*. Journal of Nematology. 2016;**48**(4):248

[136] Seo HJ, Park AR, Kim S, Yeon J, Yu NH, Ha S, et al. Biological control of root-knot nematodes by organic acid-producing *Lactobacillus brevis* wikim 0069 isolated from kimchi. The Plant Pathology Journal. 2019;**35**(6):662

[137] Conrath U. Systemic acquired resistance. Plant Signaling & Behavior. 2006;**1**(4):179-184

[138] Hammerschmidt R. Introduction: Definitions and some history. In: Induced Resistance for Plant Defense. United States: Wiley; 2014. pp. 1-10

[139] El-Deeb TS, Bakkar SM, Eltoony L, Zakhary MM, Kamel AA, Nafee AM, et al. The adipokine chemerin and fetuin-a serum levels in type 2 diabetes mellitus: Relation to obesity and inflammatory markers. The Egyptian Journal of Immunology. 2018;**25**(1):191-202

[140] Ryals JA, Neuenschwander UH, Willits MG, Molina A, Steiner HY, Hunt MD. Systemic acquired resistance. The Plant Cell. 1996;**8**(10):1809

[141] Vallad GE, Goodman RM. Systemic acquired resistance and induced systemic resistance in conventional agriculture. Crop Science. 2004;**44**(6):1920-1934

[142] Malamy J, Sanchez-Casas P, Hennig J, Guo A, Klessig F. Dissection of the salicylic acid signaling pathway in tobacco. American Phytopathological Society. 1996;**9**:474-482

[143] Chałańska A, Łabanowski G, Maciorowski R. Control efficacy of selected natural products against chrysanthemum foliar nematode–*Aphelenchoides ritzemabosi* (Schwartz, 1911) Steiner & Buhrer, 1932 Ocena skutecznościdziałanianie którychsubstancjiipreparatówpochodz enianaturalnego w zwalczaniu węgorkachryzantemowca–Aphelenchoidesritzemabosi (Schwartz, 1911) Steiner i Buhrer, 1932. Progress in Plant Protection. 2013;**53**(3):563-567